Environment Impact Assessment:
Precept & Practice

Environment Impact Assessment:
Precept & Practice

Editors

K. M. Baharul Islam
Professor
Centre of Excellence in Public Policy and Government
Indian Institute of Management Kashipur
Uttarakhand, India

****

Zafar Mahfooz Nomani
Faculty of Law
Aligarh Muslim University
Aligarh, Uttar Pradesh, India

CRC Press is an imprint of the
Taylor & Francis Group, an **informa** business

First published 2021
by CRC Press
2 Park Square, Milton Park, Abingdon, Oxon, OX14 4RN

and by CRC Press
6000 Broken Sound Parkway NW, Suite 300, Boca Raton, FL 33487-2742

© 2021 selection and editorial matter, K. M. Baharul Islam and Zafar Mahfooz Nomani; individual chapters, the contributors

CRC Press is an imprint of Informa UK Limited

The right of K. M. Baharul Islam and Zafar Mahfooz Nomani to be identified as the authors of the editorial material, and of the authors for their individual chapters, has been asserted in accordance with sections 77 and 78 of the Copyright, Designs and Patents Act 1988.

Reasonable efforts have been made to publish reliable data and information, but the author and publisher cannot assume responsibility for the validity of all materials or the consequences of their use. The authors and publishers have attempted to trace the copyright holders of all material reproduced in this publication and apologize to copyright holders if permission to publish in this form has not been obtained. If any copyright material has not been acknowledged please write and let us know so we may rectify in any future reprint.

All rights reserved. No part of this book may be reprinted or reproduced or utilised in any form or by any electronic, mechanical, or other means, now known or hereafter invented, including photocopying and recording, or in any information storage or retrieval system, without permission in writing from the publishers.

For permission to photocopy or use material electronically from this work, access www.copyright.com or contact the Copyright Clearance Center, Inc. (CCC), 222 Rosewood Drive, Danvers, MA 01923, 978-750-8400. For works that are not available on CCC please contact mpkbookspermissions@tandf.co.uk

Trademark notice: Product or corporate names may be trademarks or registered trademarks, and are used only for identification and explanation without intent to infringe.

Print edition not for sale in South Asia (India, Sri Lanka, Nepal, Bangladesh, Pakistan or Bhutan).

British Library Cataloguing-in-Publication Data
A catalogue record for this book is available from the British Library

Library of Congress Cataloging-in-Publication Data
A catalog record has been requested

ISBN: 978-1-032-05582-4 (hbk)
ISBN: 978-1-003-19820-8 (ebk)

Preface

Environment Impact Assessment: Precept & Practice deals with theoretical, practical, managerial and legal issues in multidisciplinary holism to suit environmental planning and governance. After great pains human race has understood that development at the cost of environmental degradation is a very dangerous practice which is putting very existence of this planet in great peril and most of the times the gains are nothing compared to potential hazards, and this is precisely what this book tries to view in a holistic manner. Environmental Impact Assessment is considered a single most required tool to measure all developmental goals which have equitable regime for ecosystem governance. Multidisciplinary treatment of issues impacting our lives in sectors as diverse as farming to urban planning to industrialization to legal frameworks regulating these regimes are major topics discussed with a holistic approach to manage them as per laid out Environmental Impact frameworks is a target which this work tried to achieve. The work is laced with polemical issues in dexterous detail to cater erudite demand of environmental planners besides fulfilling the void of curriculum and pedagogic requirements of technical universities, environment management experts and legal studies. Theoretical treatment of subject which is laced with deeper observation and real life examples is what will make this work a popular reference tool, for all those having a keen interest in intriguing subject of Environmental Impact Assessment.

Editors

Contents

Section I:
Environment Impact Assessment, Sustainability & Security

Chapter 1: Genuine Savings Rates in India: An Indicator of Measuring Environmental Sustainability   1–28
M. Balasubramanian

Chapter 2: Environmental Impact of Climate Change on Agricultural Sustainability, Food and Livelihood Security: A South Asian Perspective   29–52
Dr. Azim B. Pathan

Chapter 3: Effect of Eco-Literacy, Consumer Effectiveness and Perceived Seriousness on Consumer Environmental Attitude: A CFA Approach   53–78
Chirag Malik & Neeraj Singhal

Section II:
Environment Impact Assessment & Technological Innovations

Chapter 4: Water Policy in India: Building Blocks for Synergy with Science, Technology and Innovation   79–108
Venkatesh Dutta, Karunesh Kumar Shukla, Subhash Chander

Chapter 5: Use of Geospatial Technology in Environmental Impact Assessment   109–126
Ekwal Imam, Orus Ilyas & M. Mukhtyar Hussain

Chapter 6: Environmental Risk Assessment Regulation of Genetically Modified Organisms   127–150
Dr. Faizanur Rahman

Section III:
Environment Impact Assessment Climate Change & Information Management

Chapter 7: Need for Environmental Information Policy: An Analysis of Environmental Impact Assessment for Proposed Integrated Steel Complex Site, Halakundi Village, Karnataka, India 151–174
Prof. (Dr.) K.M. Baharul Islam, Archan Mitra & Dr. Asif Khan

Chapter 8: Communication for Climate Change and Control in India 175–196
Benoy Krishna Hazra & Maitree Shee

Chapter 9: Of Climate Change & The Calamitous Events of Kashmir Ecology (16th C. Onwords): An Historical Analysis 197–224
Mumtaz Ahmad Numani

Section IV:
Environment Impact Assessment & Indian Laws & Policies

Chapter 10: Environment Impact Assessment in India: Constitutional Perspective 225–242
Jaspal Singh & Varinder Singh

Chapter 11: Environment Impact Assessment Principles Under Land and Heritage Conservation Laws: An Enviro-Legal Analysis 243–266
Md. Zafar Mahfooz Nomani

Chapter 12: Environmental Impact Assessment in India: An Analysis of Law and Judicial Trends in Contemporary Perspective 267–290
Anis Ahmad

Section V:
Environment Impact Assessment & Sectoral Planning

Chapter 13: Environmental Impact Analysis: A Socio-Legal Study of Kol Dam in Himachal Pradesh   291–308
Dr. Kailash Thakur & Dr. Harish Thakur

Chapter 14: Sustainable Mining and Closure Policy Regulations and Practice: A Case Study of Coal Mining in Meghalaya   309–338
Ali Reja Osmani

Chapter 15: Environmental Impact Assessment of Mining In India: A Review of Legal and Institutional Mechanism   339–352
Aijaj Ahmed Raj & Zubair Ahmed

Chapter 1

Genuine Savings Rates in India: An Indicator of Measuring Environmental Sustainability*

M. Balasubramanian

Abstract: *GDP's current role poses a number of problems. A major issue is that it interprets every expense as positive and does not distinguish welfare enhancing activity from welfare-reducing activity. Nations, need indicators that measure progress towards achieving their goals – economic, social, and environmental. The traditional measures of investment and saving are relied upon measures based on income. The natural environment is omitted from the accounting algorithm and thus, depreciated physical capital is included, whereas, depletion of environmental resource was excluded. The World Bank attempted to make the inclusion of natural and human capital into the saving measurement, and has developed a measure known as "genuine saving' which comprises physical, human and natural capital. This paper has calculation of genuine saving rates in India and policy implications of the measures. The paper suggests that Indian approach is needed to appropriately address sustainability issues and to incorporate natural capital in national accounting.*

Keywords: *Gross Domestic Product, Welfare-Reducing Activity, Genuine Savings Rates, Natural and Human Capital, Environmental Sustainability*

1.1 INTRODUCTION

In conventional national income accounts, net savings is obtained by deducting only depreciation of physical capital from gross saving[1].

---

*This paper has presented international Conference on " Environment, Technology and Sustainable Development: Promises and Challenges in the 21st Century, held on 2nd to 4th March, 2014 organized by ABV-Indian Institute of Information

It is well understood that GNP and GDP offer an incomplete answer to the sustainability question as they omit non-marketed public goods such as environmental quality (Nordhaus, 2000). The genuine savings measure proposed by World Bank (1999). Achieving sustainable development necessarily entails creating and maintaining wealth. Given the centrally of savings and investment in economic theory, it is surprising that the effects of depleting natural resources and degrading the environment have not, until recently, been considered in measurements of national savings (Hamilton and Clemens 1999). The true rate of saving in nation after due account is taken of the depletion of natural resources and the damages caused by pollution. The new estimates of genuine saving feature broader coverage of natural resources, improved data and methods of calculation, and significant enhancement in the treatment of human resources (World Bank 1997). Pearce and Atkinson (1993) introduced a measure of sustainable development based on a net saving criterion and Hamilton (1994) has called this measure genuine savings and provided an extended framework for a more consistent treatment of natural asset loss. Genuine savings is derived as follows. A measure of gross domestic product and ‚green' net national product (NNP) in an economy with natural resources and environmental assets is (see Hartwick, 1993; Maler, 1991; Hamilton, 1994; Atkinson *et al.*1997). The main objective of the paper is analyse genuine saving calculation in India. The plan of the paper is as follows. In section 1.1 details discussion of genuine saving. In section 2 review of literature. Section 3 extends the theory and methodology of the paper. In section 4 use data for the period 1970-2008 to study the genuine saving calculation. Section 5 results and discussion and conclusion and policy implication presented in final section.

1.1.1 Genuine Savings

Genuine saving provides a much border indicator of sustainability by valuing changes in natural resources, environmental quality, and

---

Technology and Management at Gwalior. Author is gratefully acknowledged conference participants for them comments and suggestion

[1] Pearce and Atkinson (1993), Hamilton (1994, 1997, 1998), Weitzman (1996), Hartwick (1997), World Bank (1997, 2006), Atkinson et al (1998), Dasgupta and Maler (2000), Hamilton and Clemens (1999), Arrow et al (2003), Neumayer (2004, 2006)

human capital, in addition to the traditional measure of changes in produced assets provided by net saving (World Bank 2006).

Genuine saving is useful to policy makers not only as an indicator of sustainability, but as a means of presenting resource and environmental issues within a framework familiar to finance and development planning ministers. It underlines the need to boost domestic savings, and hence, the need for sound macroeconomic policies and it highlights the fiscal aspects of environment and resource management, since collecting resource royalties and charging pollution taxes are basic ways to both raise development finance and ensure efficient use of the environment. (World Bank, 2006).

Genuine savings measure seems useful as it calls for such good things as net investment in physical capital, investment in education, and subtraction from GDP of net value of natural resource depletion and the net value of environmental damage. Genuine savings indicator which is one of the best known methods to express different aspects of sustainable development in monetary terms. It is computed from the figure for net domestic savings (assumed to comprise net additions to, or investment in, human capital minus depletion of energy, mineral, and forest resources, and damage from CO_2 emission). The notion of genuine saving is presented briefly and informally in Hamilton (1994) and Perace, Hamilton and Atkinson (1996). Genuine savings have some conceptual issue and data limitations are also discussed by the World Bank[2] Fig 1.1 Show genuine saving of world counties.

An interest in the relationship among national income, wealth and welfare has revived in the recent years, driven in large part of concern about the long-run consequences of natural resource depletion and environmental degradation (Dasgupta 2003). Pearce and Atkinson (1993) provided one of the earliest suggestions for an indicator of weak sustainability: an adjusted national savings measure that accounts for the depletion of natural resources and the environment. The underplaying concept of sustainability is „weak' because it does not take account of any thresholds in critical natural asset or limits to sustainability of natural and produced asset. Hamilton (1996) developed welfare

---

[2] See Friend (2000), Hamilton Clemens (1999), World Bank (1997, 1999), Neumayer (2004, 2006, 2007), Brown et al (2003)

measures to accounts for living and non-living resources, resource discoveries and different varieties of pollutants in a similar extended national accounting framework – the corresponding saving measures, equivalent to that of Pearce and Atkinson (1993). Figure 1.2 and 1.3 respectively global percapita ecological footprint[3] by components and total wealth by types of asset in India for 2005.

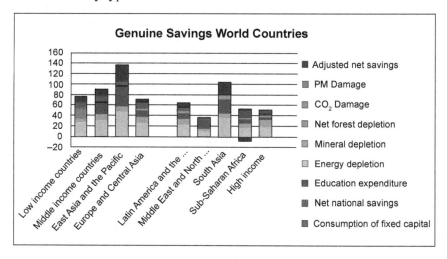

Fig.1.1: Genuine saving World Countries

Source: Author's calculation

---

[3] Ecological footprint and biocapacity of nations began in 1997. Global footprint network initiated is National footprint accounts program in 2003, with the most recent edition issued in 2001. NFAs constitute an accounting framework quantifying the annual supply of, and demand for key ecosystem services by means of two measures Ecological Footprint: a measure of the demand populations and activities place on the biosphere in a given year, given the prevailing technology and resources management of that year. Biocapacity: a measure of the amount of biologically productive land and sea area available to provide the ecosystem services that humanity consumes-our-ecological budget or nature's regenerative capacity. India's ecological foot print presented Table 4.

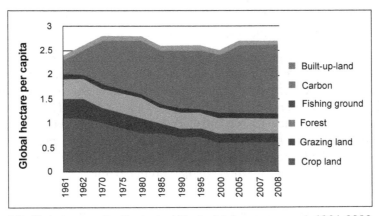

Fig 1.2: Global percapita Ecological Footprint, by component, 1961-2008

Source: Author's calculation

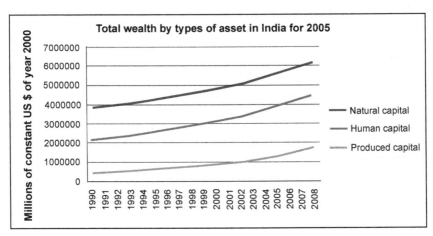

Fig.1.3: Total wealth by types of asset in India for 2005

Source: Author's calculation

1.2 LITERATURE

An extended net or genuine saving has a central place in any portfolio of indicates purporting to measure the sustainability of development appears now to be firmly established (UNECE 2007; Stiglitz, Sen and Fitoussi 2009; World Bank 2010). Pearce and Atkinson (1993) were among the first to posit a practical linkage between sustainable development and a measure of national wealth that was expanded to include natural assets.

Pearce and Atkinson proposed that this was in turn a question of maintaining total wealth and this could be measured by savings rates adjusted to reflect depletion and environmental degradation. Atkinson *et al.* (1997) and Hamilton and Clemens (1999) have updated both the theoretical argument linking savings and sustainability and the empirical estimation of adjusted net savings rates – dubbed "genuine" saving to distinguish it from traditional national accounting measures of net saving – for a wide range of countries.

Dasgupta and Maler (2000) and Asheim and Weitzman (2001) have also established further theoretical foundations to this focus on genuine saving and more broadly, comprehensive wealth accounting[4]. The link to sustainability here is that an economy should avoid negative genuine saving if development is to be sustained, if future decline in well-being are to be circumvented. Asheim (2009) shows that to use GNNI or GS for local welfare comparisons, it is sufficient to establish that GNNI growth or positive GS indicate welfare improvement, whereas for global welfare comparisons one must establish that per capita GNNI is positively related to per capita welfare. For this result to hold the economies compared must have the same technology and when estimating per capita welfare. Vincent (2001) a panel data model for 1973-97 using the GNNI and GS for 13 Latin American countries is used to test two hypotheses: GNNI should predict the present value of future consumption, and GS should predict the difference between the present value of future consumption and current consumption. Hamilton (2005) equates GS with the present value of future consumption changes and tests this using a cross-section model. The GS measures include non-renewable resources and round wood. They exclude the emissions of CO_2 and education expenditures. Four measures of saving are used, gross, net, GS and Malthusian savings (including population growth). World Bank (2011) has been a leader

---

[4] Dasgupta and Maler (2000), for example, show that net investment is equal to the change in social welfare in a non-optimising framework where a resource allocation mechanism is used to specify initial capital stocks can be mapped to future stocks and flows in this economy.

in the estimation of comprehensive wealth in sustainability analysis. The authors have estimated the (shadow) values of some components of comprehensive wealth, namely, the value of natural, human and reproducible capital. The World Bank Adjusted Net Saving data have the resource curse literature (e.g. Stoever, 2011; Barbier, 2010; Van de Ploeg and Poelhekke, 2010; Van de Ploege 2010; Dietz et al., 2007). Barbier (2010) exclude CO_2 damage from the Adjusted Net Saving calculation since it suffers from methodological flaws and data are missing for most countries for the pre-2000 period.

Mota et al (2010) have estimated two measures of weak sustainability from 1991 to 2005, the adjustments proposed are around 15% of GNI, being the environmental adjustments –depletion of forest resources and cost of air emissions of the magnitude of 7% of Gross national income. Particular matter, followed by SO_2, is the biggest contributor to total damages from air emissions accounting on average for more than half of total costs. As a % of GNI, the damages from air emissions have been decreasing. From 1990 to 2005 the best estimate is that the cost of air emissions in Portugal averages 8% of GNI with a decreasing trend, and the high estimate average 11% and the low 4% of GNI. The critical elements added by the green national accounting literature are recognition of natural resources as factors of production and of environmental amenities as sources of welfare. This is a shift that will require far more attention to be paid to environmental protection, full employment, social equity, better product quality and durability, and greater resource use efficiently (*i.e.*, reducing resource intensity per dollar of GDP). These changes are clearly within our grasp, and are underway in several countries and regions. Alternative measures of progress, like Genuine Progress Indicator (GPI), are useful to help chart and guide the course if appropriately used and understood (Kubiszewski et al 2013).

1.3 THEORY AND METHODOLOGY

1.1.2 Theoretical Framework:
Pearce and Atkinson (1993) made a first attack on the problem of measuring sustainable development by employing basic institution concerning asset sustainability. They argued that sustainability can be equated to non-declining values of all assets including natural

resources. The consequence of the conceptualization is that changes in asset values, measured by net savings should signal weather an economy is on a sustainable path. More recent theoretical work on saving had firmly established the linkage between savings, social welfare and sustainable development. Weitzman (1976) in a constant population, discounted utilitarian context with liner utility and no technological progress, showed that the current net national income, NNI, measured as the sum of consumption and not investment, is proportional to the current maximum welfare attainable along the optimal path. Hamilton and Clemens (1999) tackle the problem for an optimal economy, Dasgupta and Malar (2000) for non-optimal economies with suitable definition with (shadow prices). Asheim and Weitzman (2001) show that growth in real NNP (where prices are deflated) by a Divisia index of consumption prices) indicates the change in the social welfare in the economy

Genuine savings is defined as:

$$G = \sum p_i K_i \quad \text{----------------} \quad (1)$$

Here the K_i are the stocks of asset in the economy, and the p_i are their shadow prices. The expression says that genuine saving is measured as the change in real wealth. To measure sustainability it is important that genuine savings span as wide a range of assets as possible, including assets with negative shadow prices such as pollution stocks. In principle changes in the stock of produced, human, natural, social and institutional capital should all be measured in savings – in practice there are data and conceptual problems associated with the measurement of assets like social capital. The basic theoretical insight of Hamilton and Clemens (1999) is to show that genuine savings G, utility U, social welfare V, marginal utility of consumption λ, pure rate of time preference ρ are related as follows:

$$V = \int U(C,...), e^{-\rho(s-t)} ds \quad \text{--------------------} \quad (2)$$

$$G = \lambda^{-1} \frac{dV}{dt}$$

This says that social welfare is equal to the present value of utility and that genuine savings is equal to the instantaneous change in social welfare measure in dollars. The utility function can include consumption C and any other set of good and bad to which people attribute value. Hamilton and Clemens (1999) show that negative levels of genuine savings must imply that future levels of utility over some period of time are lower than current levels – that is negative genuine savings implies unsustainability. Sustainability literature dating back to Pearce et. al (1989) looks at the question of strong versus weak sustainability. Weak sustainability assumes that there are no fundamental constraints on sustainability.

If however, some amount of nature must be conserved in order to sustain utility – the strong sustainability assumption – then these saving models need to be modified to incorporate shadow price of the sustainability constraints. The formal approach weak and strong sustainability problem has been explored in the „Hartwick Rule' literature. Dasgupta and Heal (1979) and Hamilton (1995) show if that is elasticity of substitution between produced capital and natural resources less than 1, than the Hartiwick Rule is not feasible – eventually production and consumption must fall, implying that the economy is not substitutable under the rule.

Arrow, Dasgupta and Karl Maler intertemporal welfare for imperfect economics, (Dasgupta and Maler 2000; Arrow et.al, 2003, 2004). The wealth of an economy, w at time t can be expressed as the sum of a comprehensive se of assets K (human made capital) H (human capital) and S (natural capital), evaluated at their shadow prices

$$W_t = K_t K_t + \mu_t H_t + \lambda_t S_t \quad \text{---------------------} \quad (3)$$

constant population and exogenous movement in total factor productivity and import and export prices, the change in W, or genuine savings (GS), equates the change in social well-being when social well-being is expressed using a Ramsey-Koopmans formulation see equation (2)

$$dV_t/d_t = GS_t = k_t dK_t/dt + \mu_t dH_t/dt + \lambda_t dS_t/dt \quad \text{----------} \quad (4)$$

From (4) it follows that the value of changes in comprehensive wealth has the same sign as the corresponding change in intergenerational well-being. If $GS_t \leq 0$ the economy is deemed as unsustainable. Notice, however, that a welfare improvement at a given moment in time, $dV_t/d_t = GS_t > 0$, is weaker than long-term sustainability; i.e in an imperfect economy, the genuine savings indicator – even if perfectly estimated – is not a perfect sustainability indicator. A country seeking short-term growth at the expense of long-term viability may with a positive utility discount rate have both growth in comprehensive NNP and positive value of changes in a comprehensive vector of stocks to begin with, even through long-term sustainability is undermined (Ferreira and Moro, 2011).

1.1.3 Methodology

Pearce and Atkinson (1993) proposed a measure of „weak sustainability' which was an empirical application of the Hartwick rule[5]. Their measure has now become known as „genuine savings'. Essentially, this tests whether a country is following the Hartwick rule, by comparing the savings rate with the sum of depreciation on natural and man-made capital, all expressed as a fraction of national income.

This paper is follow standard practice Hamilton and Clemns (1999) calculate genuine savings by subtracting from gross national savings (GNS) estimates of fixed capital depreciation (D_K), depletion of natural resources (D_S), environmental degradation (D_E), and adding human capital accumulation (A_H) (see Ferreira and Mora 2011)

$$GS_t = GNS_t - D_{Kt} - D_{St} - D_{Et} - A_{Ht} \quad\text{---------------------- (5)}$$

1.1.4 Depreciation of Physical Capital:

For the calculation of physical capital stocks, several estimation procedures can be considered. Some of them, such as the derivation of capital stock from insurance values or accounting values or from direct surveys, entail enormous expenditures and face problems of limited availability and adequacy of data. Perpetual Inventory Method (PIM), are cheaper and more easily implemented since they

---

[5] The Hartwick rule requires that rents from natural resource extraction be re-invested in man-made capital to keep the total amount of capital (natural plus man-made) from decline.

require only investment data and information on the assets' service life and depreciation patterns. These methods derive capital series from the accumulation of investment series and are the most popular. The PIM is, indeed, the method adopted by most OECD countries that estimate capital stock (Bohm et al 2002; Mas, Perez, and Uriel 2000; Ward 1976). World Bank (2012) used the PIM of capital stocks. The relevant expression for computing K_t, the aggregate capital stock value in period t, is then given by

$$K_t = \sum I_{t-1} (1-\alpha) \quad\text{------------------------ (6)}$$

where I is the value of investment in constant prices and α is depreciation rate. They implicitly assume that the accumulation period (or service life) is 20 years. The depreciation pattern is geometric, with α = 5 percent assumed to be constant across countries and over time for more information (see World Bank, 2012)

1.1.5 Depreciation of natural capital

The World Bank equates the depletion of fossil fuels and minerals to current resource rent (quantity extracted times the difference between price and average total cost of extraction), using the „net price' (Repetto et al., 1989). Neumayer (2000) suggests that this can exaggerate the loss in value of reserves enormously. Moreover, net-price method captures the true value of asset depreciation only under strong assumptions of optimal management, endogenous prices and costs, and if average costs are good proxies of marginal cost (see Prrings and Vincent, 2003). Ferreira and Mora (2011) used alternative method by El Serafy (1989) this method, known as the „simple present value method' imposes no optimization on the extraction path of the resource, but assumes constant total rents and requires information on the life time of the resources. The net price and simple present value methods is that they are able to capture only the use value of the resource. For many market resources, such as fossil fuels and metals and minerals, this may be appropriate, but for other resources, such as forests, non-use values (*e.g*, the intrinsic value that people may attach to them or the option value of extending society's set of future options) (see Ferria and Moro 2011). Measuring depreciation of natural capital is empirically difficult many of these effects will be to resources with no or imperfect market values, such as biodiversity, water quality and wilderness areas (Henley et al 1999). This paper is followed by World Bank calculation, on the well-established economic principle that asset values should be measured as the present discounted value of economic profits over the life of the resources. This value, for a particular country and resources, is given by the following expression:

$$V = \frac{\sum' \pi_i q_i}{(1+r)(i-t)} \quad \text{----------------- (7)}$$

where $\pi_i q_i$ is the economic profit or total rent at time $_i$ (π_i denoting unit rent and q_i denoting production), r is the social discount rate, and T is the lifetime of the resource.

1.1.6 Environmental Degradation

The adjustment for environmental degradation consists in deducting the change in the environmental component E (i.e., air net emissions), valued at society's marginal willingness to pay (WTP) to reduce emissions (Hamilton, 1996; Dasgupta 2001); Atkinson and Hamilton, 2007). This WTP should reflect the present value of future impacts arising from current emissions (Ferreira and Moro 2011)

$$D_{Et} = WTP_t * (\Delta E_t / \Delta_t) \quad \text{----------------- (8)}$$

The WTP could be based on marginal damage cost or marginal abatement cost. These pollutants (*e.g.* noise, odoru) are neither a stock nor affect other stocks. They cause damage and thus, reduce the welfare of the population affected, but the damage (largely) ceases with exposure to the pollutant. Adjustments for their damages do not appear explicitly in genuine savings (and thus we do not consider them) but are included in green NNP (see Pezzey et al., 2006). In adjustment for CO_2 emissions, the World Bank charges global damages (a global marginal social cost of $20 per metric ton of carbon (Fankhauser 1994, 1995). Kumar and Parikh (2001a) showed that under doubled carbon dioxide concentration levels in the latter half of twenty first century the gross domestic product would decline by 1.4 to 3 percentage points under various climate change scenarios.

1.1.7 Human capital:

The process of calculating genuine savings is, in one of broadening the traditional definition of that constitutes an asset. Perhaps the most important of the additions to the asset base is the knowledge, experience, and skills embodied in a nation's populace, its human capital[6]. Human capital is more difficult to measure directly. Klenow and Rodriguez-Clare (1997), the amount of human capital per worker is defined as *exp* (rT), where r is the appropriate rate of interest, assumed to be 8.5 per cent per annum as in Arrow et al (2012), and t is the average number of years of educational attainment. The stock of human capital is the human capital per worker multiplied by number of workers. The shadow price of a unit of human capital is calculated as the total wage bill divided by the total stock of human capital. Using this method requires state-level of educational attainment and an assumed rate of return on human-capital.

1.4 DATA

All data for the analysis – Gross Domestic Product, gross saving, consumption of fixed capital[7], and depletion of natural resources (energy, minerals and net forest depletion) – are taken directly from the World Development Indicators (World Bank 2013). Total wealth, employed is derived using Perpetual Inventory Model for produced capital stock estimates, present values of mineral and energy rents, and present value of forestry, fishing and agricultural rents, all measured in constant US $ of year 2000, providing the basic estimates – these are the same total wealth data employed in

---

[6] Human capital (education, skills, tacit knowledge, health) this category is embodied in people. As teachers are painfully aware, human capital is not transferable without cost from one person to another. Education, skills, and health are ends as well as means. They have intrinsic worth, but are also of indirect value (investment in human capital raises person's productivity) (see World Bank, 2012)

[7] Ferreira et al (2003) use estimated figures for consumption of fixed capital derived from the Perpetual Inventory Model used to estimate total stocks of production capital.

(World Bank, 2012). For metals and minerals, the reported proven and probable reserves from the World Development Indicators (WDI). An extraction estimates for each commodity are obtained from the World Bank (2006) data appendix. This paper obtain the total cubic meters of commercially available forests from the *Global Forest Resources Assessment* (Food and Agriculture Organization, 2006) for 1990 and 2000 and impute commercially available forest cover linearly for intermediate years. Forest are valued not only for the wood that can be extracted from them, but also for recreation, erosion control, water filtration and habitat services they provide (Arrow *et al* 2012). World Bank estimates of annual non-timber forest benefits per hectare, one for developing countries and one for more industrialized nations. The rate of return on human capital and the level of education attainment for India apply a value of 0.085. The average educational attainment, measured in years, for the adult population in India increased by an incredible 12 per cent from 1995 to 2000 (Arrow et al 2012). India CO_2 emissions data obtained from the 2012 World Development Indicators[8]. India damage from climate change bears 5 percent calculated by (Arrow *et al* 2012). For PM_{10}, due to the absence of alternative reliable and detailed data on emission and marginal external costs for the period 1995-2000 in India. The World Bank estimates population-weighted average levels of PM_{10} for all cities with a population in excess of 100,000 in each country. Particulate emission damage is calculated as the WTP to reduce the risk of mortality attributable to PM_{10} (Pandey *et al.*, 2005).

1.5 RESULTS

Global wealth reached $673,593 billion in 2005 (table 1). Table Intangible capital was the largest single component in all regions, and its share increase in importance with rising income, from 57 percent of total wealth in low-income countries to 81 percent in high-income countries. We see a symmetrical decline in the importance of natural capital as income rises, from 30 percent in low-income countries to 2 percent in high-income countries. But does this apparent relationship between the composition of wealth

---

[8] Available online at hppt://web.worldbank.org/.

and income, seen when comparing different regions at a point in time, really hold for a given income group as it develops over time[9]. Total wealth grew rapidly over the decade, outpacing population growth so that produced capital and human capital respectively China are 6159399 and 8727850 among the BRICS countries. Russia is highest contribution of natural capital 6856502 compare than other countries seen the table 2. Table 2 clearly demonstrates that total wealth by recently, emerging developing economies will becoming a strong wealth of nations around the world. Figure 2 shows the national accounting flows used to estimate adjusted net saving in India. Adjusted net saving is equal to gross savings minus consumption of fixed capital, plus education expenditure, minus energy depletion, mineral depletion, net forest depletion, carbon dioxide damage, and particulate matter (PM damage) (World Bank 2012). The dynamics are captured by adjusted net saving or genuine savings, defined gross national savings adjusted for the annual changes in the volume of all forms of capital. Since wealth changes through saving and investment, ANS measures the annual change in a country's national wealth (World Bank, 2012).

India's gross national saving has increased the highest in the world. In fact, the savings rates of many of the advanced countries and some of the Asian emerging market economies witnessed a decline during this period. India's savings rate declined sharply in 2008, as it did in many other countries, in the aftermath of the global financial crisis, but recovered, to some extent, in 2009 Figure 1.3. Consumption of fixed capital[10] has increased in 1990s around 10% of Gross national income because of increasing consumption of goods and services clearly shows Figure 1.4. Green accounting methods to the measurement of net savings appears in Pearce and

---

[9] Analysis of wealth accounts in this chapter is based on data for 124 countries for which wealth accounts are available for 1995, 2000, and 2005, as described in annex 1. An additional 28 countries for which wealth accounts are available only from 2000 are not included in this analysis (World Bank 2012)

[10] Consumption of fixed capital or depreciation is calculated by superimposing a pattern of decline in value over this time. This concept refers to that part of gross product which is required to replace fixed capital used up in the production process during an accounting period. Consumption of fixed capital represents the reduction in the value of fixed assets used up in production process during the accounting period resulting from physical deterioration, normal obsolescence or normal accidental damage (CSO, 2010)

Atkinson (1993), who combine published estimates of depletion and degradation with standard national accounting data to calculate true savings for 20 countries (Hamilton and Clemens, 1999). Figure 1.4 In fig – the net savings rate for India over 1970-2008 was driven by the increasing trends from 8.65% to 29.68%. In per capita terms, expenditure at constant prices shows a sharper decline. So education expenditure has stagnated and even declined in spite of the rhetoric on the part of the Government of India. This is also borne out by the trend in the share of public education expenditure in GNI. The share was around from 2.47 per cent in 1970 and 3.17 per cent in 2008. As a proportion of total revenue expenditure, a decline in present decade is observed-at the turn of the century more than 14 per cent was spend on education, but around 12 per cent is spent in the present decade in GDP. The World Bank's estimate of investment in human capital is the most questionable. It equals UNESCO estimates of current operating expenditures on education. One problem is that this is a purely gross measure, which makes no allowance for losses in human capital (Ferreira and Vincent, 2005). The process of calculating genuine is, in essence, one of broadening the traditional definition of what constitutes an asset. Perhaps the most important of the additions to the asset base in the knowledge, experience, and skills embodied in a nation's populace, its human capital (Hamilton and Clemens 1999). As a proportion of GDP the situation looks worse: the ratio of education expenditure to GDP had declined from more than 4 per cent to 3.5 per cent from 1990 to mid nineties and recovered to over four per cent around the year 2000. But since that date it has been decreasing again. So, at the macro-level, education has been struggling to maintain its importance (De and Tanuka Endow, 2008).

Forest depletion is simply valued as the stumpage value[11] (price minus average logging cost) of the volume of commercial timber and fuel wood harvested in excess of natural growth in commercially valuable wood mass for that year. Harvest rates by country are as given FAO (1994). The foregoing description of the valuation of forest depletion suggests that the calculations are quite rough. It should also be obvious that the value calculated pertain only to

---

[11] Stumpage rates come from World Bank data, Openshaw and Feinstein (1989), Kellenberg (1995), and others.

commercial exploitations, so that the values of biodiversity, carbon sequestration, and other uses are not captured (Hamilton and Clemens, 1999). Fig 7 it is evident that the net forest depletion in India from 1970 to 2008, 1.39 per cent to 0.78 percent. Figure 1.7 Among the natural capital accounts, stock accounts for mineral and energy resources are complied most regularly. Energy depletion[12] was difficulties, World Bank assigned dollar values to the stocks of the main energy resources (oil, gas, and coal)[13] and to the stocks of 10 metals and minerals (bauxite, copper, gold, iron ore, lead, nickel, phosphate rock, silver, tin, and zinc for all the countries that have production figures). In fig 8 energy depletion (% of GNI) in India was 4.86 as of 2008, while its lowest value was 0.48 in 1970. Mineral depletion [14](% of GNI) in India was 1.42 as of 2008, while its lowest value was 0.15 in 1970 seen Fig 1.9. Environmental degradation in most detail because it is evident that it is playing a dominating role in our computation of genuine savings. Environmental degradation are arising from CO_2, SO_2, NO_x and PM_{10} emissions into the atmosphere. A recent study (Ghoshal and Bhattacharya, 2007) estimated carbon dioxide (CO_2) emissions for India based on fuel usage patterns in different states in India. Fig 1.10 is the CO_2 damage (% of GNI) in India was 1.16 as of 2008, while its lowest value was 0.52 in 1970. Pollution damage can enter green national accounts in different ways. Although damage to produced assets (the damage to building materials caused by acid rain, for example) is in principle included in depreciation figures, in practice most statistical systems are not detailed enough to pick this up. The effects of pollution on output (damaged crops, lost production owing to morbidity) are usually not broken out explicitly,

---

[12] Energy depletion is equal to the product of unit resource rents and the physical quantities of energy extracted. It covers crude oil, natural gas, and coal.
[13] Coal is subdivided into two groups: hard coal (anthracite and bituminous) and soft coal (lignite and subbituminous)
[14] Mineral depletion is equal to the product of unit resource rents and the physical quantities of mineral extracted. It refers to bauxite, copper, iron, lead, nickel, phosphate, tin, zinc, gold and silver.

but because they are reflected implicitly in the standard national accounts, there is no need to adjust savings measures in this regard.[15]

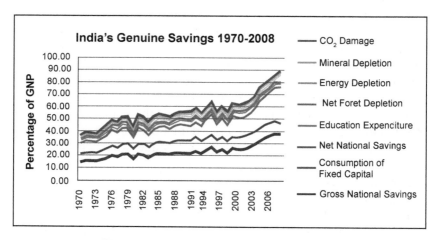

Fig. 1.4: India's Genuine saving 1970-2008

Source: Author's calculation

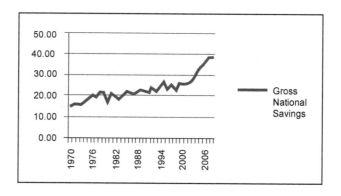

Fig.1.5: Gross National Savings

---

[15] However, if the productive capacity of an asset, such as soil fertility, is damaged by pollution, then the loss in asset value should be deducted from savings (see Hamilton and Clemens, 1999).

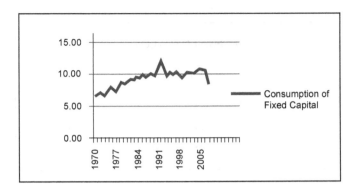

Fig. 1.6: Consumption of Fixed capital

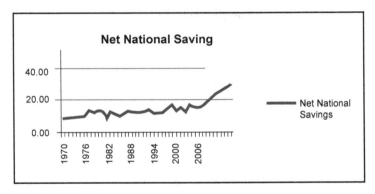

Fig. 1.7: Net National Saving

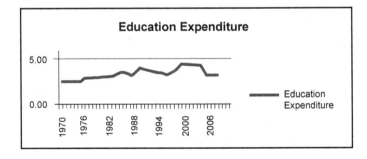

Fig.1.8: Education Expenditure.

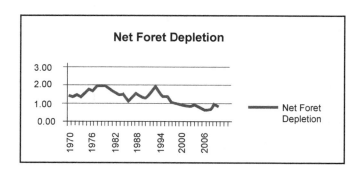

Fig.1.9: Net Foret Depletion

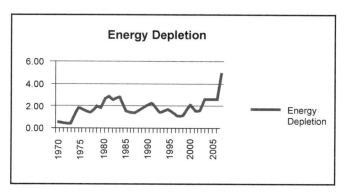

Fig. 1.10: Energy Depletion

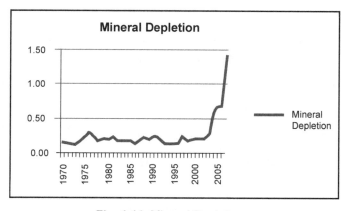

Fig. 1.11: Mineral Depletion

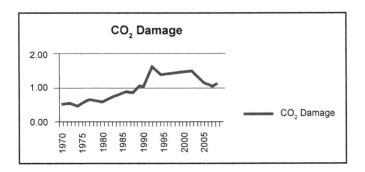

Fig. 1.12: CO_2 Damage

1.6 CONCLUSION AND IMPLICATION

The picture of genuine saving rates is not complete, however, just as standard measures of savings ignore asset consumption in natural resource capital such as forests, so can it be argued that they ignore investment in one of a nation's most valuable asset: its people (Hamilton and Clemens, 1999). This paper computes genuine savings indicators for India. The collection of data from World Bank and Indian sources and was more difficult than envisaged, some of the data were confidential and not accessible. The genuine saving estimates use the present value method for the valuation of the depreciation of tradable natural resources. This method is endorsed by national and international statistical agencies (UN, 1993) and preferred to the net-price method used by the World Bank. Human capital accumulation using the returns to education, this method is favoured by the literature over the education. Expend the valuation environmental degradation by including estimates of SO_2 and NO_x external cost. Existing studies of genuine investment have been compiled primarily on the basis of the World Bank's methodology. Data limitations have also been a major constraint in estimating comprehensive investment measures. There is a need for better investment estimates, particularly for natural and human capital. Adjustments for natural resource depletion for underground water resources and biodiversity for instance remain a difficult task given the limitations of available methods for valuing these services (Dasgupta and Shikla Gupta, 2010). In terms of further research there are obvious refinements that can be envisioned, including treating soil degradation and expanding the country coverage of data on the marginal social costs of pollution emissions. The latter is particularly important for rapidly growing countries, as countries grow, they end to urbanise, and these urban areas tend to develop problem levels of pollution (Hamilton and Clemens, 1999). This paper also suggests need more study for ecological foot and biocapacity in India for measurement of sustainable development.

Genuine savings measures suggest a series of policy questions that are key to sustaining development. While natural resource exports boost foreign savings and therefore the overall savings effort, the analysis of genuine savings suggests a further question: to what do exhaustible resources boost the rate of genuine savings? The policy question for natural resource management is therefore the extent to which stronger resource policies (royalty regime, tenure) can boost the rate of genuine savings. Similarly, reducing pollution emissions to socially optimal levels will boost the value of genuine savings. The policy issue with respect to pollution is the extent to which more optimal pollution control polices can increase the rate of genuine savings. The policy implications of measuring genuine savings are quite direct, persistently negative rates of genuine must lead, eventually, to declining well being. For policymakers, linking sustainable development to rates of genuine savings means that there are many possible interventions to increase sustainability, from the macroeconomic to the purely environmental (Hamilton and Clemens 1999). Development to be sustainable if future utility does not decrease below the current level, i.e., utility is non-decreasing (Pezzey 2004). Genuine savings presents so far the best attempt at measuring weak sustainability with considerable scope for future developments and improvements (Dietz and Neumayer 2004)

Table 1: Total wealth by types of asset and income group

	3597	57	13	30
Low Income				
Lower Middle Income	58023	51	24	25
Upper Middle Income	47183	69	16	15

Contd...

Higher Income	551964	81	17	2
OECD				
World	673593	77	18	5

Source: World Bank (2012)

Note: Figures are based on the set of countries for which wealth accounts are available from 1995 to 2005. High-income oil exports are not shown.

Table 2: Total wealth by BRICS countries for 2005

	Brazil	Russia	India	China	South Africa
Produced Capital	1464091	1335387	1736341	6159399	321335
Human Capital	4559539	2135477	2843220	8727850	1042978
Natural Capital	1390147	6856502	1584404	5072761	481755

Source: World Bank (2012)

Table 3: India's wealth estimates for 2005 (US $ per capita)

Timber	195
Non-timber forest resources	17
Protected areas	145
Crop land	1391
Pasture land	602
Natural capital	2704
Produced capital + urban land	1980
Intangible capital	5961
Total Wealth	**10539**

Source: World Bank (2012)

Table 4: Ecological Footprint in India
(population in million and global hectare per person)

Year/Indicators	1996	1999	2001	2003	2005	2007	2008
Population Million	949.9	999.2	1033.4	1065.5	1103.4	1134	1190.9
Cropland	0.46	0.28	0.34	0.34	0.4	0.39	0.37
Grazing Land	0.05	0	0	0	0.01	0	0
Forestland	0.12	0.11	0.01	0.08	0.1	0.12	0.12
Fishing ground	0.01	0.03	0.05	0.04	0.01	0.02	0.02
Carbon	0.31		0.03	0.26	0.33	0.33	0.31
Built-upland	0.01	0.06	0.04	0.04	0.04	0.05	0.05
Total Ecological foot print	1.06	0.77	0.8	0.8	0.9	0.91	0.87
Cropland	2.64	0.28	0.29	0.29	0.31	0.4	0.38
Grazing land	4.25	0	0	0	0.1	0	0
Forestland	2.3	0.31	0.02	0.02	0.02	0.02	0.02
Fishing groud		0.04	0.03	0.03	0.04	0.03	0.03
Built-upland						0.05	0.05
Biocapacity	0.74	0.68	0.04	0.4	0.4	0.51	0.48
Ecological deficit	-0.32	0.09	-0.4	-0.4	-0.5	-0.4	-0.39

Source: World Wildlife Fund (2012)

REFERENCES

1 Arrow, K., Dasgupta, P., Mäler, K.-G., 2003. Evaluating projects and assessing sustainable development in imperfect economies. Environmental and Resource Economics 26, 647-685
2 Arrow, K.J., P. Dasgupta, L. Goulder, G. Daily, P.R. Ehrlich, G.M. Heal, S. Levin, K.-G.

3   M¨aler, S. Schneider, D.A. Starrett, and B. Walker (2004), „Are we consuming too much?', *Journal of Economic Perspectives* **18**(1): 147–172.
4   Arrow J Kenneth, Partha Dasgupta, Lawrence H Goulder, Kevin J Mumford and Dirsten Oleson (2012) Sustainability and the Measurement of Wealth, Environmental and Development Economics, 17: 317-353.
5   Asheim, G. and Weitzman, M. (2001). Does NNP growth indicate welfare improvement?
6   *Economics Letters* 73, pp. 233-239.Atkinson and Hamilton, 2007
7   Atkinson, Giles (1998) Discussion at the meeting on 'Alternatives to economic statistics as indicators of national well-being'. *Journal of the Royal Statistical Society: series A (statistics in society)*, 161 (3).
8   Atkinson, Giles and Dubourg, Richard and Hamilton, Kirk and Munasinghe, Mohan (1997) Measuring sustainable development: macroeconomics and the environment. Edward Elgar, Cheltenham, UK.
9   Barbier, E., 2010. Corruption and the Political Economy of Resource-Based evelopment: A Comparison of Asia and Sub-Saharan Africa. Environmental & Resource Economics 46, 511–537.
10 Bohm, B., A. Gleiss, M. Wagner and D. Ziegler (2002), „Disaggregated capital stock
11 estimation for Austria – methods, concepts and results', Applied Economics 34, 23–37.
12 Costanza, R., Hart, M., Posner, S., Talberth, J., 2009. Beyond GDP: The Need for New Measures of Progress
13 Friend, A.M., 2000. Roots of green accounting in the classical and neoclassical schools. In: Proops, Simon (Ed.), Greening the Accounts. Edward Elgar, Cheltenham, UK.
14 Dasgupta, Partha S. 2003. *Human Well-Being and the Natural Environment*. Rev. ed.
15 Oxford: Oxford University Press
16 Dasgupta P and Heal G (1979) Economic Theory and Exhaustible Resources, Cambridge University Press, Cambridge.
17 Dasgupta, P. and K.-G. M¨aler (2000), „Net national product, wealth, and social wellbeing', *Environment and Development Economics* **5**(1): 69–93.
18 Dasgupta Purnamita & Shikha Gupta, 2008. "Measuring Sustainability with Macroeconomic Data for India," Macroeconomics Working Papers 22149, East Asian Bureau of Economic Research.De Anuradha and Tanuka Endow (2008) Public Expenditure on Education in India: Recent trends and Outcomes, Research Consortium on Education Outcomes and Poverty
19 Dietz, S., Neumayer, E., 2004. Genuine savings: a critical analysis of its policy-guiding value. International Journal of Environment and Sustainable Development 3, 276–292.
20 Dietz, S., Neumayer, E., 2007. Weak and strong sustainability in the SEEA: concepts and measurement. Ecological Economics 61 (4), 617–626.
21 El Serafy, S., 1989. The proper calculation of income from depletable natural resources. In: Yusuf, A., El Serafy, S., Lutz, E. (Eds.), Environmental Accounting for Sustainable Development: A UNDP-World Bank Symposium. World Bank, Washington DC, pp. 10 -18.

22 Fankhauser, S., 1994. The economic costs of global warming damage: a survey.
23 Global Environmental Change 4, 301-309.
24 Fankhauser, S., 1995. Valuing Climate Change e The Economics of the Greenhouse, first ed. EarthScan, London
25 Ferreira Susana and Mirko Moro (2011) Constructing genuine savings indicators for Ireland, 1995-2005, Journal of Environmental Management, 92: pp 542-553.
26 Ferreira, S. and J. Vincent (2005). Genuine Savings: Leading Indicator of Sustainable Development? *Economic Development and Cultural Change*.
27 Ferreira, S., K. Hamilton and J. Vincent (2003). Comprehensive Wealth and Future
28 Consumption. (mimeo). Washington: The World Bank.
29 Food and Agriculture Organization (2006), „Global Forest Resources Assessment 2005', United Nations FAO Forestry Paper No. 147. Rome: FAO.
30 Friend, A.M., 2000. Roots of green accounting in the classical and neoclassical schools. In: Proops, Simon (Ed.), Greening the Accounts. Edward Elgar, Cheltenham, UK.
31 Ghoshal and Bhattacharya, 2007
32 Hamilton, Kirk. 1994. "Green Adjustments to GDP." *Resources Policy* 20(3):155–68.,
33 Hamilton 1996. "Pollution and Pollution Abatement in the National Accounts." Review of Income and Wealth 42(1, March):13–33.
34 Hamilton 1997. "Defining Income and Assessing Sustainability." World Bank, Environment Department, Washington, D.C. Processed.1998
35 Hamilton, K., and J.M. Hartwick, 2005. Investing Exhaustible Resource Rents and the Path of Consumption. Canadian J. of Economics 38:2, 615-21, May 2005.
36 Hamilton, K., 1996. Pollution and pollution abatement in the national accounts. Review of Income and Wealth 42, 13 - 33.
37 Hamilton, K., Atkinson, G., 1996. Air pollution and green accounts. Energy Policy 24, 675-684
38 Hamilton, K. and M. Clemens (1999), „Genuine savings rates in developing countries', World Bank Economic Review 13(2): 333–356.
39 Hartwick, John M. 1977. "Intergenerational Equity and the Investing of Rents from
40 Exhaustible Resources." American Economic Review 67 (December): 972–74.
41 Hartwick, John M., and Anja Hagemann. 1993. "Economic Depreciation of Mineral Stocks and the Contribution of El Serafy." In Toward Improved Accounting for the Environment, ed. Ernst Lutz. Washington, DC: World Bank.
42 Hanley, N., Moffatt, I., Faichney, R., Wilson, M., 1999. Measuring sustainability:a time series of indicators for Scotland. Ecological Economics 28, 55-73.

43 Klenow, P.J. and A. Rodr´ıguez-Clare (2005), „Externalities and growth', in P. Aghion and S. Durlauf (eds), Handbook of Economic Growth, Amsterdam: North Holland.
44 Kellenberg, John. 1995. "Accounting for Natural Resources: Ecuador, 1971–1990." Ph.D. diss. Department of Geography and Environmental Engineering, Johns Hopkins University,Baltimore, Md.
45 Kubiszewski Ida Kubiszewski, Robert Costanza, Carol Franco, Philop Lawn, John Talberth, Tim Jackson and Camille Aylmer (2013) Beyond GDP: Measuring and Achieving Global Genuine Progress, Ecological Economics, 93: 57-68.
46 Kumar, K.S. Kavi, and J. Parikh (2001a), "Socio-economic Impacts of Climate Change on Indian Agriculture", International Review of Environmental Strategies, 2(2): 277-293.
47 Mas, Matilde, Francisco Perez and Ezequiel Uriel (2000), „Estimation of the stock of capital in Spain,' Review of Income and Wealth 46:1, 103–16.
48 Mota, R., T. Domingos, V. Martins (2010). Analysis of genuine saving and potential green net national income: Portugal, 1990–2005. Ecological Economics 69, 1934-1942.
49 Neumayer, E. (2000), „Resource accounting in measures of unsustainability: challenging the World Bank's conclusions', Environmental and Resource Economics, 15, 257–78
50 Nordhaus, W.D. and J. Boyer (2000), Warming the World: Economic Models of Global Warming, Cambridge, MA: MIT Press.
51 Openshaw, Keith, and Charles Feinstein. 1989. "Fuelwood Stumpage: Financing Renewable Energy for the World's Other Half." Policy Research Working Paper 270. World Bank, Policy Research Department, Washington, D.C. Processed
52 Pandey, K., Bolt, K., Deichman, U., Hamilton, K., Ostro, B., Wheeler, D., 2005. The Human Cost of Air Pollution: New Estimates for Developing Countries. Development Research Group and Environment Department, World Bank, Washington
53 Pearce, D.W. and G. Atkinson (1993), „Capital theory and the measurement of sustainable development: an indicator of weak sustainability', Ecological Economics 8: 103–108.
54 Pearce D W Markandya A and Barbier E (1989) Blueprint for a Green Economy, Earthscan Publications, London.
55 Pezzey, J., 2004. One-sided sustainability tests with amenities, and changes in technology, trade and population. Journal of Environment Economics and Management 48, 613–631.
56 Pezzey, J.C.V., Hanley, N., Turner, K., Tinch, D., 2006. Comparing augmented sustainability measures for Scotland: is there a mismatch? Ecological
57 Economics 57, 60-74.
58 Repetto, R.,W. Magrath, M.Wells, C. Beer, and F. Rossini (1989),Wasting Assets: Natural Resources in the National Income Accounts, Washington, DC: World Resources institute
59 Stiglitz, J.E., Sen, A., Fitoussi, J.P., 2010. Mismeasuring Our Lives: Why GDP Doesn't Add Up. The New Press, New York.

60 Stoever, J., 2011. On Comprehensive Wealth, Institutional Quality and Sustainable Development – Quantifying the E_ect of Institutional Quality on Sustainability. Journal of Economic Behavior & Organization
61 Van der Ploeg, F., Poelhekke, S., 2010. The Pungent Smell of „Red Herrings': Subsoil Assets, Rents, Volatility and the Resource Curse. Journal of Environmental Economics and Management 60, 44–55.
62 Vincent, J. 2001. Are greener national accounts better? Center for International Development at Harvard University Working Paper No. 63.
63 Ward, M. (1976), The Measurement of Capital. The Methodology of Capital Stock Estimates in OECD Countries, OECD, Paris.
64 Weitzman, M. L. 1976. "On the Welfare Significance of National Product in a Dynamic Economy". Quarterly Journal of Economics 90(1):156–62.
65 World Bank (1997), Expanding the Measure of Wealth: Indicators of Environmentally
66 Sustainable Development, Washington, DC: World Bank.
67 World Bank (2006), Where is the Wealth of Nations? Washington, DC: World Bank.
68 World Bank (2011) The Changing Wealth of Nations: Measuring Sustainable Development in the New Millennium, Washington, DC: World Bank.
69 World Bank, 1999. World Development Indicators 1999. World Bank, Washington, DC.
70 World Bank (2012) The Changing Wealth of Nations Measuring Sustainable Development in the New Millennium, Environment and Development.

Chapter 2

Environmental Impact of Climate Change on Agricultural Sustainability, Food and Livelihood Security: A South Asian Perspective

Dr. Azim B. Pathan

Abstract: In this chapter, author has attempted to examine national and international legal instruments and policies relating to climate change and sustainable development through the lens of livelihood challenges for farmers. Researcher has also attempted to explore the farmers concern in different national and international legal instruments, declarations and policies. This paper also finds out an impact of climate change on farmers and agricultural sector, which is the backbone of economy for many South Asian countries. Despite of the efforts of different international organizations, still problems like, climate change induced livelihood challenges for farmers, food security problems etc. in South Asian countries including Bangladesh, China and India, are persisting. Besides, in this paper researcher has vigorously argued that concern of farmers is only the missing link between climate change induced problems and sustainable development. Sustainable development will be a reality when livelihood challenges of farmers and especially agricultural sector will be addressed in the climate change related legal and policy documents. This paper takes a broader view and explores the multiple effects that climate change can have on farmers, food production and food security. It also tries to explore an adaptation and mitigation measures, especially in the agricultural sector with reference to farmers of South Asian countries.

Keywords: *Farmer, Policy, International Legal Instrument, Sustainable Development, Climate Change, Livelihood, South Asian Countries etc.*

2.1 INTRODUCTION

We are talking too much about climate change at global arena through International Conferences, Negotiations, Declarations, Policies and Plans, but are not able to pin point, who are ultimately getting affected due to menacing problem of climate change? Whose interest ultimately is at stake and threat, due to historical fact of more emission caused by developed countries including United States of America (U.S.A.)? Whether International Legal Instruments and Policies are capable enough to sustain interests and human rights of poor people, who are dependent on agriculture for their livelihood? These are some major questions which should be addressed properly for getting a suitable model for bridging the gap between climate change related international legal instruments and sustainable development. It is significant to note that situation of agriculture and crop productivity of developing countries, especially South Asian countries, including Bangladesh, China and India have been continuously getting affected due to climate change-induced floods, drought and unseasoned monsoon, consequently it has direct impact on the livelihood of farmers and agriculture sector. Moreover, it is significant to take into consideration the basic idea of „sustainable development' which was reflected through different international conferences and declarations. Brundtland Commission Report of 1987 has clearly stated about the sustainable development as the development which meets the needs of present generation without compromising the needs and aspirations of future generations. The principle related to sustainable development which was reflected through United Nations Conference on Environment and Development, 1992 in Rio-de-Janeiro, had mainly emphasized on the inclusive development, taking into consideration society, economy and environment, based on the Brundtland Commission Report. Despite the efforts of United Nations Organization especially to prevent the menace of climate change in 1992, through framing the United Nations Framework Convention on Climate Change (UNFCCC) problem still persists. Conversely, different issues are cropped up due to rising global warming such as climate change-induced livelihood challenges for farmers and threatened crop productivity in South Asian countries. International legal instruments such as UNFCCC, 1992 does not address the issue of climate change induced livelihood challenges for farmers. The most severe impact of climate change is being felt by vulnerable populations of South Asian countries who have contributed least to the problem as compared to developed countries such as U.S.A. and European countries. Climate change is an inevitable and urgent global challenge with long-term implications for the sustainable development of all countries including South Asian countries. Climate change is expected to impact nearly every aspect of life; natural disasters are becoming more frequent and intense, agricultural conditions are changing, and diseases are becoming more prevalent.

A key challenge in responding to climate change is the increasing number of events of floods. From 1999 to 2008, floods affected almost 1 billion people in Asia.[1] The corresponding figures were about 4 million in Europe, 28 million in the Americas and 22 million in Africa. For instance, the 2010 flood in Pakistan affected more individuals than the combined impacts of the Indian Ocean tsunami (2004), the Kashmir earthquake (2005) and the Haiti earthquake (2010). Flash floods in the Himalayas are estimated to cause the loss of at least 5,000 lives every year.[2] Some of the most direct and immediate impacts of climate change are on agriculture. Rising sea levels displacing coastal farming communities, declining water supplies, and shifting weather patterns raising the specter of drought and crop failures are just some of the effects already being experienced across the globe. As a result, agricultural systems, people, and institutions will come under increasing pressure and stress. Law and legal systems are significant social institutions with critical roles to play in shaping a more sustainable future for nations as well as individual farmers. Climate change

---

[1] UN Women Watch, 2009, in *Women at the Frontline of Climate Change- Gender Risk and Hopes, UNEP,* Center for International Climate and Environmental Research, also available at <http://www.grida.no/files/publications/women-and-climate-change/rra_gender_screen.pdf>, Last visited on 08.10.2014.

[2] *Ibid.*

presents new challenges to the ideals of sustainable agriculture and sustainable development and ultimately those, whose livelihood is depended on agriculture.[3]

Climate change has emerged as one of the most important environmental issues ever to confront humanity. It is seriously affecting the ecosystem worldwide. The effect of climate change and global warming is also being felt in South Asian countries including India. It is taking place due to the increasing concentration of carbon dioxide in the atmosphere. Global warming which is the rising in temperature of earth leading to climatic changes, which are taking place due to the emission of carbon dioxide, greenhouse gases[4] and fume discharges from the Chemical Industries. It is fact that our everyday activities may be leading to changes in the earth's atmosphere that have the potential to alter the planet's heat and radiation balance. The last decade of the 20th Century has been reported to be the warmest decade[5]. Climate change affects mostly rural people and those who are dependent on agriculture. It hurt poor countries and poor communities most. It is significant to note that the effect of climate change is mainly felt by developing countries including South Asian countries. Climate change is already exacerbating chronic environmental threats, and ecosystem losses are restraining livelihood opportunities, especially for farmers who are dependent on agriculture. Farmers and agricultural sector have become more vulnerable to the effect of climate change. This is apparent fact in the developing countries, especially in South Asian countries including India, that climate change and global warming has direct impact on agriculture, consequently upon the livelihood of farmers. Global warming has direct effect in the form of floods, droughts, loss of crops, less agriculture productivity, farmers' suicides due to failure of crops etc. It is affecting more to the people of rural area whose livelihood is dependent on the farming and agriculture.

---

[3] Neil D. Hamilton, Farming an Uncertain Climate Future: What Cop 15 means for Agriculture, U. Ill. L. Rev. (2011), at p. 341.

[4] Greenhouse gases means the Carbon dioxide, Chlorofluorocarbons, etc. which are released in atmosphere by anthropogenic activities.

[5] *DainikJagran*, Varanasi, Aug. 26, 2002, p. 10.

2.2 WARSAW OUTCOME & FARMERS' CONCERN

With the success of the Millennium Development Goals(MDGs) the members of the UN met together to formulate the SDGs to carry the achievements of the MDGs for next 15 years i.e. from 20165-2030 with the seventeen goals to completely achieve that what was stared in 2000 with the introduction of the eight MDGs. The second of the seventeen proposed SDGs is "End hunger, achieve food security and improved nutrition, and promote sustainable agriculture". It aims at by 2030 that the world should achieve double of the agricultural productivity and the incomes of small-scale food producers, particularly women, indigenous peoples, family farmers, pastoralists and fishers, including through secure and equal access to land, other productive resources and inputs, knowledge, financial services, markets, and opportunities for value addition and non-farm employment.[6] Increase investment, including through enhanced international cooperation, in rural infrastructure, agricultural research and extension services, technology development, and plant and livestock gene banks to enhance agricultural productive capacity in developing countries, in particular in least developed countries.[7]

The 19th Session of the Conference of Parties (COP 19) held at Warsaw, Poland laid the foundation for 2015 Climate Agreement by keeping governments on track. Warsaw has set a pathway for governments to work on a draft text of a new Global Climate Agreement which may be placed on table in December of this year in Climate Change Conference (COP 20) at Lima, Peru. The death and destruction brought by the Philippines Storm helped to highlight the question of climate justice[8]. The "loss and damage mechanism" under the treaty witnessed difference of opinion between developing countries and developed countries. The United States and European

---

[6] Sustainable development knowledge platform
 (https://sustainabledevelopment.un.org/?page=view&nr=164&type=230&menu=2 059)
[7] *Ibid.*
[8] The spirit of climate justice is the understanding that the urgent action needed to prevent climate change must be based on community-led solutions and the well-being of local communities, Indigenous Peoples and the global poor including farmers, as well as biodiversity and intact ecosystems.

Union opposed the mechanism proposed by developing countries, fearing new financial claims. The mechanism relating to "loss and damage" which was suggested by developing countries especially South Asian countries to tackle climate change induced problems like decreased agriculture crops and livelihood challenges for farmers could not become reality because of indifferent attitude of developed countries to accept obligation for loss and damage by climate change.

2.3 CLIMATE CHANGE IMPACT ON BANGLADESH'S AGRICULTURE AND FARMERS

It is predicted that climate change could have devastating impact on agriculture. Agriculture is a key economic driver in Bangladesh, accounting for nearly 20 percent of the GDP and 65 percent of the labor force. The performance of this sector has considerable influence on overall growth, the trade balance, and the level and structure of poverty and malnutrition. The United Nations Environment Programme estimates a one-meter rise in sea level would inundate 17,000 square kilometers of Bangladesh's land, over ten percent of its total land mass. This is a serious threat to a country like Bangladesh. Moreover, much of the rural population, especially the poor, is reliant on the agriculture as a critical source of livelihoods and employment. The impacts of climate change could affect agriculture in Bangladesh in many ways:

1. The predicted sea-level rise will threaten valuable coastal agricultural land, particularly in low-lying areas.
2. Biodiversity would be reduced in some of the most fragile environments, such as sunder bans and tropical forests.
3. Climate unpredictability will make planning of farm operations more difficult.

The effects of these impacts will threaten food security for the most vulnerable people of Bangladesh. The country's agriculture sector is already under stress from lack of productivity and population growth. Any further attempt to increase productivity will likely to add pressure to available land and water resources. In Bangladesh, a rise in sea level of 1 meter would lead to the inundation of 15% to

18% of the country's land mass, and by the year 2050, a total of 30 million people.[9] Bangladesh has already achieved one of the key Millennium Development Goals (MDG) - gender parity in primary and secondary schooling. The country is on track to achieve most of the MDG goals, even the difficult ones like infant and maternal mortality by 2015. Agriculture is the largest employment sector in Bangladesh. As of 2016, it employs 47% of the total labour force and comprises 16% of the country's GDP. The Government of Bangladesh signed a US$10 million grant agreement on 30 September 2013 with the World Bank to introduce solar irrigation pumps for farmers.[10] The Solar Irrigation Project will enable installation of more than 1,500 solar powered irrigation pumps covering more than 65,000 bighas of land for rice cultivation. The project will be financed by the Bangladesh Climate Change Resilience Fund (BCCRF).[11] However, the predicted adverse impacts due to global warming could reverse the recent economic and social gains. The progress towards achieving the MDGs, such as eradicating poverty, combating communicable diseases and ensuring environmental sustainability could be in jeopardy.

2.3.1 Climate Change Strategy and Action Plan (2009)

The Government of Bangladesh (GOB) took the initiative to prepare the Climate Change Strategy and Action Plan 2009 ("the BCCSAP 2009") to carry forward and coordinate climate change activities in the country. BCCSAP 2009 has identified six broad areas of actions (i) food security, social protection and health; (ii) comprehensive disaster management; (iii) infrastructure; (iv) research and knowledge management; (v) mitigation and low carbon development; and (vi) capacity building and institutional strengthening.[12] In BCCSAP 2009 one missing link is that of farmers' livelihood which is dependent on agriculture. A Climate

---

[9] United Nations Development Programme, 2007.
[10] https://www.bccrf-bd.org/NewsEvents.html , Last visited on 17.10.2016.
[11] *Ibid.*
[12] Government of the People's Republic of Bangladesh, Bangladesh Climate Change Strategy and Action Plan, 2009,
<http://www.moef.gov.bd/climate_change_strategy2009.pdf>, Last visited on 22.08.2014.

Change Unit has been established at the Department of Environment and Forests to coordinate the overall activities on climate change in the country. The implementation of the BCCSAP 2009 will be financed through GOB's own resources, including Climate Change Trust Funds and bilateral and multilateral support. US$ 10 billion will be required to implement BCCSAP 2009 during five years. Even though strong action plan was devised by the Government of Bangladesh for combating climate change, still there is gap between climate change adaptation resilience and sustainable development, including the social progress, avoiding livelihood challenges of farmers.

2.4 CLIMATE CHANGE RESILIENCE FUND (BCCRF) & VISION STATEMENT[13]

The Government and development partners in Bangladesh agreed on a fourfold vision statement for BCCRF. By 2020 the BCCRF will be completely a government led, owned and managed and would work as a sustainable climate change financing mechanism, aimed at developing capacity and resilience of the country to meet the challenges of climate change. BCCRF will support the implementation of the BCCSAP through an institutional framework by:

1. Providing a platform for coordination of BCCRF stakeholders and acting as a catalytic agent for wider coordination.
2. Serving as a climate fund, which also brings innovation, harmonization and added value to the GoB's climate change initiatives.
3. Serving as a financing mechanism to bring global climate change funding to Bangladesh.
4. Supporting implementation of prioritized, results-oriented climate change interventions that deliver sustainable outcomes particularly targeting the least resilient.

---

[13] Ibid.

2.5 CHINA AND CLIMATE CHANGE IMPACT ON AGRICULTURE AND FARMERS

China's agricultural GDP has grown at an average of 5 percent annually over the last three decades, and while this figure is less than the 10 percent annual growth in total GDP, it is nonetheless respectable. Within the agricultural sector, significant structural changes have taken place. Commodities' share of the total value of agricultural output fell from 82 percent in 1970 to less than 50 percent after 2006.[14] Researchers have long pointed out that rural people are particularly vulnerable to climate change, especially in the case of extreme weather events such as droughts and hailstorms.[15] In the dry regions of western China such as Ningxia Province, a reduction in rainfall and an increase in the incidence of drought in recent years have been shown to affect local farming activities and livelihood.[16]

2.5.1 China's Policies and Actions for Addressing Climate Change, 2013

The National Development and Reform Commission have organized China's Policies and Actions for Addressing Climate Change the compilation of the National Plan for Addressing Climate Change (2013-2020).17 After an overall analysis of the trends and impacts of climate change in China, as well as the current situations and challenges in addressing climate change, the National Development and Reform Commission proposed the main target, major tasks and safeguarding measures for addressing climate change by 2020.18 Additionally, it has outlined the general framework for addressing climate change in China. All provinces (including the autonomous

---

[14] Climate Change and China's Agricultural Sector-An Overview of Impacts, Adaptation and Mitigation, International Center for Trade and Sustainable Development, 2010, available at
<http://www.agritrade.org/events/do cuments/ClimateChangeChina_final_web.pdf>, visited on 20th October, 2014.

[15] Ibid.

[16] Ibid.

[17] *See*<http://www.ccchina.gov.cn/archiver/ccchinacn/UpFile/Files/Default/20131108091218654702.pdf>, Last visited on 23.08.2014.

[18] Ibid.

regions and municipalities directly under the central government) have taken active steps in carrying out the formulation of mid and long-term plans for addressing climate change at the provincial level.19 It is significant to note here that National Plan for Addressing Climate Change (2013-2020) does not specifically take into consideration livelihood challenges for farmers. It also does not talk about how to deal with the loss of agricultural crops due to climate change, which is major issue in developing countries including China.

2.6 FIVE-YEAR CLIMATE SMART AGRICULTURE PROJECT

In 2014, the National Development and Reform Commission has invested over 20 billion yuan within the central government's budget to support the construction of production bases for agricultural products such as grain and cotton, and to strengthen field projects based on small-sized farmland hydrological projects to improve disaster prevention and mitigation capabilities.[20][21] The China government seeing the then present scenario aimed working on to promote and expand the Dryland Agriculture Technology and develop new methods and techniques to support dry land water-saving agriculture in north, northeast and northwest China.[22] The Ministry of Agriculture and Global Environment Facility (GEF) has jointly invested in a *Five-year Climate Smart Agriculture Project* in major grain production bases. The project aims at increasing the adaptation of farming to climate change and promoting energy saving and emissions reduction in agriculture.[23] In 2015, The

---

[19] Ibid.

[20] The National Development and Reform Commission November 2014- China's Policies and Actions on Climate Change 2014

[21] http://en.ccchina.gov.cn/archiver/ccchinaen/UpFile/Files/Default/20141126133727 751798.pdf

[22] Ibid.

[23] Ibid.

Ministry of Agriculture (MOA) has worked in conjunction with China Meteorological Administration to organize work such as:[24]

1. Strengthening Guidance and Services to Win Summer Grain Harvest
2. Fighting Spring Drought and Flood to Ensure Spring Seeding in Northeast, and
3. Enhancing Services and Fighting Disasters to Win Autumn Grain Harvest.
4. The main aim of the National Plan for Addressing Climate Change (2014-2020), is to achieve low-carbon development in saving energy, streamlining energy structure, adjusting industrial structure, ecological construction and environmental protection, engage in international negotiations on climate change in an active and constructive way, continue to promote bilateral and multilateral dialogues, communication and pragmatic cooperation on climate change, and make greater contribution to the protection of global climate environment. Even these goals are aimed at but at the same time the agricultural sector gets no attention and it keep on fighting against the climate changes.

2.7 IMPACT OF CLIMATE CHANGE ON INDIAN AGRICULTURE AND FARMERS

India's agriculture is more dependent on monsoon from the ancient periods. Any change in monsoon trend drastically affects agriculture. Even the increasing temperature is affecting the Indian agriculture. In the Indo-Gangetic Plain, these pre-monsoon changes will primarily affect the wheat crop. In the states of Jharkhand, Odisha and Chhattisgarh alone, rice production losses during severe droughts average about 40% of total production, with an estimated value of $800 million. It has direct impact on the farmers specially those who hold fragmented pieces of lands.

---

[24] The National Development and Reform Commission November 2015- China's Policies and Actions on Climate Change 2015

Recent studies done at the Indian Agricultural Research Institute indicate the possibility of loss of 4-5 million tons in wheat production in future with every rise of 1-degree Celsius temperature throughout the growing period. Rice production is slated to decrease by almost a tonne/hectare if the temperature goes up by 2 degree Celsius.[25] In Rajasthan, a 2 degree Celsius rise in temperature was estimated to reduce production of Pearl Millet by 10-15%. If maximum and minimum temperature rises by 3 degrees Celsius and 3.5 degree Celsius respectively, then Soybean yields in Madhya Pradesh will decline by 5% compared to 1998. Agriculture will be worst affected in the coastal regions of Gujarat and Maharashtra, as fertile areas are vulnerable to inundation and Salinization.[26]

2.7.1 A Scanning of Farmers Concern in Twelfth Five Year Plan (2012-2017) of Indian Government

It is significant to note that the Twelfth Five Year Plan (2012-2017) of Government of India recognizes the inclusive development. The key strategies in the Twelfth Plan have been identified as:

1. Economic Empowerment;
2. Social and Physical Infrastructure;
3. Enabling Legislations;
4. Women's Participation in Governance; and
5. Inclusiveness of all categories of vulnerable groups

Twelfth Five Year plan states that there are inclusive growth concerns in all aspects. It also suggests climate change adaptation strategies will be made a part of all ongoing poverty reduction and development policies, including Disaster Risk Reduction (DRR) planning and implementation at local, national and regional level, country's Nation Adaptation Programmes of Actions (NAPAs); and in numerous climate change related funds that are in the process of being established. This Five Year Plan also identified the climate change induced problem such as low agriculture productivity. But

---

[25] Anupama Mahato, Climate Change and Its Impacts on Agriculture, International Journal of Scientific and Research Publications, Volume 4, Issue 4, April 2014, available at <http://www.ijsrp.org/research-paper-0414/ijsrp-p2833.pdf>, visited on 21st October, 2014.

[26] *Ibid.*

the main concerns of farmers and their livelihood challenges and mainly farmers' suicides due to due loans and crop failures have not been adequately address by Twelfth Five Year Plan.The major impacts of climate change will be on rain fed or un-irrigated crops, which are cultivated on nearly 60 percent of cropland. A temperature rises by 0.5C in winter temperature is projected to reduce rain fed wheat yield by 0.45 tonnes per hectare. Possibly there might be some improvement in yields of chickpeas, rabi maize, sorghum and millets and coconut on the west coast and less loss in potatoes, mustard and vegetables in north-western India due to reduced frost damage. Increased droughts and floods are likely to increase production variability.[27] Vulnerability to climate change is closely related to poverty, as the poor have fewer financial and technical resources. They are heavily dependent on climate-sensitive sectors such as agriculture and forestry; they often live on marginal land and their economic structures are fragile.[28]

2.7.2 Impact and vulnerability of Indian Agriculture to Climate Change

Indian agriculture today is faced with the challenge of having to adapt to the projected vagaries of climate change. It must develop mechanisms to reduce its vulnerability. The Indian Council of Agricultural Research (ICAR) has already begun research to assess the likely impact of climate change on various crops.

2.7.3 Cereal Crops

The Asia-Pacific region is likely to face the worst impacts on cereal crop yields. Loss in yields of wheat, rice and maize are estimated in the vicinity of 50%, 17%, and 6% respectively by 2050.[29] This yield loss will threaten the food security of at least 1.6 billion people in South Asia. The projected rise in temperature of 0.5°C to 1.2°C will be the major cause of grain yield reduction in most areas of South Asia.

---

[27] International Policy Digest- Climate Change and Indian Agriculture, Abhimanyu shrivastava, 22 August,2016

[28] Climate Change and Indian Agriculture, Anish Chatterjee

[29] Report, Indian Council of Agricultural Research (ICAR).

2.7.4 Wheat

India is considered to be the second largest producer of wheat and the national productivity of wheat is about 2708 kg/ha. The Northern Indian states such as Uttar Pradesh, Punjab, Haryana, Uttaranchal and Himachal Pradesh are some of the major wheat producing states. Here the impact of climate change would be profound, and only a 1°C rise in temperature could reduce wheat yield in Uttar Pradesh, Punjab and Haryana. In Haryana, night temperatures during February and March in 2003-04 were recorded 3°C above normal, and subsequently wheat production declined from 4106 kg/ha to 3937 kg/ha in this period.[30]

2.8 CONSENSUS ON CLIMATE CHANGE JUSTICE

At this juncture it is important to note here that climate change is not the concern of only one nation but it is the concern of whole world. It is reality that the developed countries like U.S.A. are not ready to reduce their industrial emission and the developing countries are also not prepared to reduce their emission up to the standard level. Even after recent Warsaw Conference on Climate Change, CoP 19 the stand of USA is still not favorable towards acceptance of binding emission cut agreement unless China and India accept the binding emission cut agreement. At the same time developed and developing countries are defending their activities within their territories on the basis of right to development. The developed countries have already polluted the environment and now they are asking the developing countries including India to reduce their industrial emission and greenhouse gases. This conflict between developed and developing countries is the main obstacle in solving the problem of global warming.

---

[30] *Ibid.*

International efforts to address this problem have been ongoing for the last two decades, with the Earth Summit at Rio in 1992 as an important launching point. Although India as a developing country does not have any commitment or responsibility at present for reducing the emissions of greenhouse gases such as CO2 that lead to global warming, pressure is increasing on India and other large, rapidly developing countries such as China and Brazil to adopt a more pro-active role. At the same time, the developed countries of the North are trying to limit the extent of their commitments for emission reduction. In this situation, the public and policy makers need to be aware of the ramifications and implications of the problem of global warming. Moreover, it is significant to take into consideration the principle 21 of the Stockholm Declaration which is the major outcome of the United Nation Conference on Human Environment held at Stockholm in 1972 and the principle 2 of Earth Summit or United Nations Conference on Environment and Development held at Rio-de-Janeiro, which provide that states have sovereign right to exploit their own resources pursuant to their own environment policies and the responsibility not to damage the environment of states beyond the limits of national jurisdiction. But these principles are not binding on the nations. This is the major hurdle in the way of determining the liability and responsibility for the problem of climate change and global warming.

It is significant to note that the 1992 Rio Conference adopted the United Nations Framework Convention on Climate Change which aims to stabilize CO2 emissions and other anthropogenic gases which accumulate in the atmosphere and trap the heat of the sun and enhance the greenhouse effect.[31] But the United Nations Framework Convention on Climate Change, 1992 deals with the relation between state and state. It does not deal with the issue of responsibility of state towards individual and specially that vulnerable group of poor people who are mostly dependent on the agriculture for their livelihood. In this background, there is legal gap in climate change regime to deal with the issue of climate change affected people especially farmers. Even recent Warsaw Climate Change Conference, 2013 held at Poland and different decisions which were taken including Decision 2/CP.19 related to „Warsaw

---

[31] United Nation Framework Convention on Climate Change, 1992, Art. 2.

international mechanism for loss and damage' does not specifically talk about how to deal with loss and damage caused to the agricultural crop productivity and livelihood to the farmers of South Asian countries due to climate change.

The Climate Convention as a tool to defend against the threat of global warming and as a means for sustainable progress has some major weaknesses. For instance, it is compromise document and sets a voluntary goal of returning to 1990 levels by 2000 and the limit have been set by Kyoto Protocol from 2008-2012. IPCC Report says that desired outcome has not come in reality. This is the major reason which dilutes the effectiveness of the Convention. The financial mechanism provided in the Convention help the developing countries in terms of finance as well as new technologies.[32] But new sophisticated technologies require appropriate infra-structure, skills and institutional arrangements which are not available in the developing countries. India lacks such human and infrastructural capacity and because of that its efforts towards the problem of climate change are not meager. The technological lag is felt even in terms of productivity, where the outcome is leading to concentration of fertilizers found in the food available, at the same time, quantitative difference is also not sufficing for the rate at which subsequent population growth is taking place. Under the Convention, the developed countries are supporting the development and enhancement of indigenous capacities and technologies of developing countries. But in reality developed countries are not providing technology and finance in adequate manner. The competitive market is burdening the developing countries in a very sophisticated manner, where now owing to the fact that the developed countries wanting to keep their market large, the technological developments passed on are either flawed or aimed at profit of the developed nations. This can be seen in various strategies being employed. For example, the current spread of Ebola virus, can be seen from two perspectives. This is like an extension of the old policy used during the times when vaccination was not available to countries being exploited. Here as well, it can be seen that the spread of this virus is pertinent in a few areas only. So, a *factum* of dependence is developing on the developed nations.

---

[32] United Nation Framework Convention on Climate Change, 1992, Art. 11.

In December 2015, the 21st Session of the Conference of the Parties (COP21/CMP1) convened in Paris, France, and adopted the Paris Agreement, a universal agreement whose aim is to keep a global temperature rise for this century well below 2 degrees Celsius and to drive efforts to limit the temperature increase even further to 1.5 degrees Celsius above pre-industrial levels.[33] It is paramount to note that economies of developing countries are highly dependent upon fossil fuels and energy intensive products. Moreover, the developing countries are using obsolete and environmentally unfriendly technology supplied by the developed countries. In such circumstance it is difficult for the country like India to take adequate measure towards the menace of climate change. At the same time developing countries have no access to advance and environment friendly technology which is protected by Intellectual Property Rights of developed nations. So the challenge is how to ensure clean technologies to the developing world. The recent trend however, is to a small extent contributing to technological advancement at a small pace though. For example, recently, as was promisingly seen in Ted Talks, Shubhendu Sharma, who was working with Toyota in India devised a methodology to make a forest grow ten times faster than normal. Here, he used various mechanisms that sped the process of growth of plants, trying to replicate the technology used in car manufacturing, creating a system allowing a multilayer forest of 300 trees to grow on an area as small as the parking spaces of six cars, for less than the price of an i-phone. He now, is an eco-entrepreneur who is also working towards soil quality improvisation.

It seems that this is what the developing nations need today, better technologies which are compatible with the conditions of the nations. The developed nations do have enough insight to have extensive farming methodologies employed. However, nations like India, Pakistan, Bangladesh and other developing South Asian nations have always had a high dependability on farmers, along with the fact that intensive farming having been the methodology employed all the time. So, keeping in mind the fact that climate changes are being rendered, the need of hour is to seek technological

---

[33] Sustainable development knowledge platform
(https://sustainabledevelopment.un.org/topics/climatechange)> visited on 17-10-2016

help in this sector and device the mechanisms which justify the existing conditions. Keeping too much in lines with the path taken up by developed nations is not the exact solution we seek.

Sustainable Development Goal 13 aims to "take urgent action to combat climate change and its impact", while acknowledging that the United Nations Framework Convention on Climate Change is the primary international, intergovernmental forum for negotiating the global response to climate change.[34] More specifically, the main targets of SDG 13 better focuses on the integration of climate change measures into national policies; the improvement of education, creating awareness and institutional capacity on climate change mitigation, adaptation, impact reduction and early warnings.[35] Moreover, the Kyoto Protocol, 1997 on global warming enjoins on industrialized countries to reduce their greenhouse emissions by 8% of 1990 emission levels. It is important to note here that USA has refused to ratify the Kyoto Protocol. This move of the USA towards the Protocol and consequently towards the menacing problem of global warming is also influencing the efforts of other developed countries.

International concerns also arise due to the inability of ratifying nations to meet their commitments. The failure of signatories to meet their obligations, combined with the lack of repercussions signals to Annex I and developing nations that the Protocol is likely to fail. What is also a considerable fallacy in these estimations is their proximity to be achieved. Even in context of developing nations like India, the radar set is too high to achieve. Change in political setups, advancements and other factors influence the existing conditions high enough to cause considerable deliberation. So, the set parameters should also incline towards achievable standards.

The lack of accurate emission baselines creates an obstacle in determining whether participating states are meeting their Protocol commitments. Some signatories may have an advantage when 1990 baselines are chosen to assess emission rate changes. In nations such as Russia, where the national economic output has declined since

---

[34] *Ibid.*
[35] *Ibid.*

1990, GHG emissions may also have decreased.[36] This situation creates "head room" or "hot air."[37] Conversely, the Protocol's use of 1990 emission levels as a baseline burdens some countries, like the United States. A further obstacle to the Protocol's effectiveness may be leakage, a phenomenon which occurs when businesses move their operations and accompanying emissions to unregulated countries to avoid the expenses of obtaining emission-reduction technology. Essentially, even if the Annex I countries reduce overall emissions, those emissions may resurface elsewhere, as non-Annex I countries enjoy the economic benefits of leakage.

In addition to this, developing countries argue that they do not have adequate resources even to meet their basic human needs and livelihood. In such circumstances climate mitigation will bring additional strain to the already fragile economies of India and other developing countries. It is apparent that developing countries have lower per capita income and their contribution to the greenhouse gases concentration is very little as compared to the industrialized nations. But in present scenario situation has little change and as fast growing developing countries per capita emission level of China and India has increased. It is the fact that developed countries are historically responsible for emission concentrations which contribute the global warming. Developing countries further argue that the socio-economic development and poverty eradication are the most important elements of environment protection and for maintaining ecological balance. The efforts are being made by the developing countries to reduce the emission of greenhouse gases but the efforts made by them are not sufficient. Moreover, at the same time they are fighting with pollution and other environmental degradation problems at their national level. Therefore, they are insisting the developed world to meet its commitments on emissions of greenhouse gases. They have also pointed out the fact that developed and the developing countries should not be brought under one and the same category as far as reduction of harmful gas emission is concerned.

---

[36] *See* Scott Barrett, Environment and Statecraft: The Strategy of Environmental Treaty-Making, (2005), at p. 382.

[37] Ibid.

The major problems which are contributing the global warming and the climate change in our country are the population growth, industrialization, massive deforestation, vehicular pollution, etc. At the same time the efforts of the Indian government towards the climate change are being influenced by the international complex scenario towards the issue of global warming.

In India there is no specific law which deals with the problem of global warming. India is such a country which is having constitutional provisions for the protection of environment. There are legislations such as Air (Prevention and Control of Pollution) Act, 1981, Forest Conservation Act, 1980, Environment Protection Act, 1986, etc. for the environment protection but these legislations are the piecemeal legislations and are general in nature. They did not specifically deal with the problem of climate change, global warming and especially in reduction of GHGs. Even though there are international efforts towards the hazard of climate change and global warming, hardly any effort was taken by the Indian government through legislations for curbing the situation of reduction of GHGs. It is apparent that the NAPCC has not made integrated efforts and co-ordinate with other policy and plans to reduce the carbon emission in India. There is no integrated policy and plans of the central government which deal with the threat of climate change.Even the judiciary has shown the concern towards the problem of climate change where Hon'ble Apex Court in the landmark case of Indian Council for Enviro-Legal Action vs. Union of India[38] observed,

"If the mere enactment of laws relating to protection of environment was to ensure a clean and pollution free environment, then India would perhaps, be the least polluted country in the world. But this is not so. There are stated to be more than 200 Centre and State Statutes which have at least some concern with Environment Protection, Either directly or indirectly. The plethora of such enactments has unfortunately not resulted in preventing environment degradation, which on contrary, has increased over the years".[39] It is apparent from the above observation of Supreme Court that by

---

[38] AIR 1996 SC 1446.
[39] AIR 1996 SC 1446.

having only enactments and laws, the problem of environment degradation and climate change cannot be solved.

2.9 CONCLUSION AND SUGGESTIONS

South Asian countries' vulnerability to future disasters is profound, principally for reasons of population and poverty. The majority of South Asian countries are low-or lower-middle income countries that already struggle to support the daily needs of their growing populations. A rising sea level and reduced production in agriculture pose the biggest threats. Kolkata, Mumbai, and Dhaka, whose greater urban areas are home to over 46 million people and rising, face the greatest risk of flood-related damage over the next century. Low-lying Bangladesh is vulnerable to flooding and cyclones in the Indian Ocean, which scientific literature suggests will grow more intense in coming decades. In 1991, a 20-foot storm surge that followed a cyclone killed nearly 140,000 people in Bangladesh and left up to 10 million homeless.

As shifting monsoons affect the amount of water available for irrigation, a reduction in the standard of living is all but guaranteed for a region that already is one of the world's most water-stressed. By mid-century, the IPCC projects that climate change will make South Asia home to the "largest numbers of food-insecure people." So it comes down to what we really need to ensure that environmental challenges are not posing a significant threat to the livelihood of each one of us, especially the farmers. It needs to be analyzed that the fallacies are corrected at ground level. Let's analyze them one by one. The first thing which we should consider is the problem of smaller land holdings, it is not seen as a significant issue but when seen through the production issue along with the effect being caused to the land in use, it is the most crucial problem. The land suffers in terms of quality, and hence the production quality goes down, moreover the produce is very small, so sometimes all it does is suffice itself for the personal consumption of the farmer's family. There is no issue to it not serving a commercial value; the problem is with the fact that the farmer has to go hunting for alternative sources of income. This big problem can be solved by simple consolidation of landholdings into bigger units and giving joint ownership to the farmers where they can employ distribution of the various works involved, also since profitable impetus has been seen through this mechanism, they can easily afford advanced ways of approaching farming and ensure that not only their personal needs are met, but also those of the market with improvised production quality.

The second problem faced is that of usage of bad methodologies in improvising production. Here in developing countries, we have high dependence on fertilizers to improve the production, we do not analyze the technology employed with respect to the need, we orient ourselves to just one dimension i.e. improving quantitative production and the mechanism fails terribly. Hence, here, what is needed is to reduce dependence on fertilizers and analyze the situation case sensitively. This means that the soil needs to be given resources to recover itself, along with ensuring that its components are restored more naturally. The technology devised by Afforestry,[40] shows that the land could be covered with layers of hay after water is supplied, to ensure that the water is effectively absorbed by the plants and not merely lost to evaporation. At the same time, it also showed that seeds were planted relatively closer than they are done, consequently it grows with more productivity. Many other alternatives were also suggested by Afforestry, and along with the current research alternatives suggested by those researches, whoare viewing this in the light of sustainable development, should be employed and better results will be ensured.

The third problem is with the adequacy of the environmental conditions available. Yes, pollution and deforestation are in true terms affecting the production capacity for the farmers. So to a small extent, we need regulated weather conditions to ensure that the natural production is favorable. Here, we are not suggesting regulating the weather condition, but what we are suggesting is to ensure that the environmental conditions are not hampered as much as possible. This can be understood in a much better way via an example, for example the plantations in Himachal Pradesh are to be studied. Here, the existing policies should be as such that they regulate the environmental conditions are not hampered to affect the natural production of the crops grown in there. This also means

---

[40] The eco-enterprise of Mr. Shubhendu Sharma, available at< http://india.ashoka.org/fellow/shubhendu-sharma>, visited on 20th October, 2014.

ensuring that if more human settlement is occurring then, the regulating authority ensures that the waste produced has proper dumping facilities available, with separation of degradable and non-degradable substances, etc. States like Chhattisgarh in India, faced rapid industrialization, especially cities like Raipur where sponge iron factories grew in numbers rapidly post state creation in 2000. This went rampant and un-regulated. The effect was seen on rice production, Chhattisgarh was called the rice bowl of India, however, the quality of rice available degraded. Moreover, being more objectively viewed from a commercial standpoint, the natural production suffered and who were the worst hit- the farmers. They felt the need to move to cities to find alternative sources of income, hence more workers in the city meant cheap labour, and this further meant more industrial undertakings coming into picture. We see this setup is very unidirectional. So the need of hour is to regulate the direction of this circle towards more enhanced production mechanism, bringing it to such a standard point, that children in future, after studies should want to come back to farming. This can only be done if the right set of technological advancements is employed and the same is conveyed to the farmers.

Moreover, one suggestion would be to expand the U.S.-India Energy Dialogue to include all nations in the region and invite more involvement and investment from American private enterprise. This collaboration could satisfy both the developing world's need for support, and the developed world's desire to promote their own renewable industries, especially farther up the production value chain.For example, the advancements in battery technology being made by Tesla may greatly improve the storage of electricity generated from renewable sources. Developing countries including South Asian countries could use these new techniques to avoid building expensive power grid systems where smaller networks of battery-equipped solar facilities would be more appropriate, such as rural areas.

The lack of a „cooperative regional framework' as well as ongoing disputes between the developed and developing countries are two of the largest factors inhibiting an effective response towards climate change induced problem such as decreased agricultural productivity in South Asian countries. Unfortunately, the cost of action will only rise if delayed. It may be a line to say that "political will" is the solution to this difficult problem of governance, but in this case it is very true. On a conclusive note, it can be said that there is a significant amount of work that needs to be done in ensuring better livelihood of farmers in South Asian Countries. The essence lies in ensuring that the intention behind principles of Sustainable Development is ensured, from the scratch. International instruments, national legal instruments do exist to mandate and understand as to what is it that we are looking to fulfill. However, in their realization to the level of implementation, there is a significant gap which needs to be first understood, then realized. It can only be carried out if the way in which it is being viewed is changed, from a commercial stand point to a more practical one. Exploitation has rendered in the past for us facing two world wars, and who suffered? So, when the consequences are for all of us to face, then why look at things with exploitative eyes, and not a more understandable one. Hence, it needs to be ensured that the underlying intention behind the devised international principle of Sustainable Development is realized.

Chapter 3

Effect of Eco-Literacy, Consumer Effectiveness and Perceived Seriousness on Consumer Environmental Attitude: A Confirmatory Factor Analysis (CFA) Approach

Chirag Malik & Neeraj Singhal

Abstract: *Purpose: To study the socio-psychological factors - eco-literacy, perceived consumer effectiveness and perceived seriousness of environmental problem - influencing the environmental attitude of the consumers of Delhi-NCR. For meeting the research objective, descriptive research method was used. A survey was administered to the residents of Delhi-NCR which was divided into four zones yielding a total response of 347 consumers. Standard scales with some relevant modification were used for measuring socio-psychographic constructs and consumer environmental attitude and then exploratory factor analysis and confirmatory factor analysis were used to find out the nature of relationship amongst these constructs. Results show that Perceived Seriousness of Environmental Problem (PSEP), Eco-literacy and Perceived Consumer Effectiveness(PCE) are three socio-psychological factors which are positively correlated to consumer environmental attitude (CEA). The research is conducted in a small area of Delhi-NCR using convenience sampling technique which is more prone to common bias error. Only three socio-psychological factors were studied in this paper whereas other socio-psychological and socio-cultural factors are yet to be studied with respect to their relationship with consumer environmental attitude. The* *demographic factors have also not been studied. Further research can be done on the moderation effect of product involvement, pricing system and product necessity on the CEA. This research can be extended over to the relationship between CEA and the willingness to buy environmentally friendly product. The outcome of the research may help the marketers, policy makers, strategists, advertisers and other stakeholders in designing their policies and promotional strategies in order to tap a new emerging market of socially responsible and environmentally friendly consumers.*

Keywords: *Consumer Environmental Attitude, Eco-Literacy, Perceived Consumer Effectiveness, Perceived Seriousness of Environmental Problem, Socio-Psychological Variables*

3.1 INTRODUCTION

Among all the industrial developmental activities in India, the protection of the environment and the sustainable development has always taken the centre stage. Most of the policies governing the industrial development have considered the protection of environment and its sustenance. Ministry of Environment & Forests, which is responsible for the environmental protection and preserve natural resources of India has come up with several policy initiatives to prevent the deterioration of the environment due to industrialization and increased consumption level. The decades of 1980s and 1990s have witnessed a large scale of exploitation of natural resources, which made the ministry to come up with more stringent legislations to control the rising pollution level. One such initiative is the Notification on Environmental Impact Assessment (EIA) of developmental projects issued on 27.1.1994 under the provisions of *Environment (Protection) Act*, 1986. Under this provision, it was made mandatory to do EIA for 30 categories of developmental projects in total till 2000.

Human, through its activities puts its impact on the environment. Generally, the impact is not positive i.e. favourable to the environmental health. Every anthropogenic activity has some impact on the environment. However, it is essential for mankind to undertake these activities for its physiological, social and psychological needs. As a result, this development ignores the important aspect of human life i.e. the environment. There is need to harmonize the human development with the environmental consciousness and concern. Environmental impact assessment (EIA) is one of the tools available with the planners to achieve the above-mentioned goal. It is the demand of the hour that the sustainability has to be considered by availing any development option. But, the role of consumer as a significant contributor to the overall deterioration of the environment quality has always been ignored by the government authorities. Talking about the strategic building of environmental laws and policies, Land use regulation and climate change policies, social impact of reframing environmental laws and policies, the consumer and/or end-user related policies are still at the initial stage of law-making.

Now the individual environmental impact can also be measured which is called carbon footprint measure. This is done using various demographic parameters and the lifestyle of an individual. Now a new term has been coined, called ecological footprint of an individual, which is more relevant in terms of assessing the environment impact of an individual's overall activities and consumption. In short we can say that ecological footprint is the environmental impact of an individual beyond its carbon foot print. The basis and rationale for the measurement of ecological footprint is to assess that how much an individual draw the resources of the planet Earth and how many planets (having the Earth's resource capacity) would be required to fulfilled the consumption demand of Earth's population with the set of individuals with their specific lifestyle.

In India the Ministry of Environment, Forest and Climate Change, has taken an initiative to educate and sensitize the people at all levels towards the problem of environmental degradation and climate change. A scheme was launched in 1983 which was known as „Environmental Education, Awareness and Training (EEAT)'. The objective of the scheme is to make the people understand the relationship of environment and humankind. Other major objective is to build capabilities and skills to protect the environment. Following are the sub-objectives of the schemes;

1. To promote environmental awareness among all sections of the society;

2. To spread environment education, especially in the non-formal system among different sections of the society;
3. To facilitate development of education/training materials and aids in the formal education sector;
4. To promote environment education through existing educational/scientific/research institutions;
5. To ensure training and manpower development for environment education, awareness and training;
6. To encourage non-governmental organizations, mass media and other concerned organizations for promoting awareness about environmental issues among the people at all levels;
7. To use different media including films, audio, visual and print,, theatre, drama, advertisements, hoarding, posters, seminars, workshops, competitions, meetings etc. for spreading messages concerning environment and awareness; and
8. To mobilize people's participation for preservation and conservation of environment.

Now the time has come to focus on measuring the impact of EEAT in terms of actions taken by the people to protect the environment in order to assess the effectiveness of the campaign. Eco-literacy has a role to play in shaping the attitude of people towards the environment but „does it converts to the actual behaviour'? The environmental impact of various population factors such as population size, population distribution, population composition along with mediating factors like technology, policy context and cultural factors were assessed by Rand Corporation, USA.

Thanks to the efforts made by the government agencies, NGOs and local environmentalists, the awareness level of the society with regard to its impact on the environment has increased. Now, the environmental concern of the people has also taken a place in overall decision making by the consumers (Manaktola and Jauhari 368). This phenomenon is known as „consumer environmentalism'. Now there is emergence of new buzz word amongst business community „corporate environmentalism', which is the result of the growth of „consumer environmentalism'.It is related to the responsibilities of the businesses to consider themselves as a part of the wider community and can influence the environmental degradation in a positive way. (Banerjee et al.148). It has been observed that most of the consumers are concerned about the environment but this concern is not manifested in their actual attitude and behaviour. However, people have become more sensitive about the environmental degradation but their ultimate action for reversing this process of environmental degradation is under observation. This can be reflected in the involvement of consumers in environmental caring activities such as (a) showing no inhibition in using recycled products, helping in the process of recycling the packaging or the used products, saving energy through using lesser energy consuming appliances, (b) acceptance of lower technical performance of the product purchased due to its better eco-performance, (c) purchase of CFC free, biodegradable and organically produced products, (d) willingness to pay premium on eco-friendly products, tendency to prefer eco-friendly service provider in the areas of hospitality and tourism (Aitken 978).

Having said that, organizations and consumers, separately have been accused of being responsible for environment degradation and ecological imbalance. Organizations use non-recyclable materials, outdated machinery or equipments causing high carbon emission, and substantial industrial waste. Consumers, on the other hand, have also been blamed for their uncontrollable consumption of goods(Banerjee 150; Menon & Menon 55).Organizations have different motives for working in the direction of protecting the environment. Long term gain through casting a green company image and tapping the market of the environmentally friendly consumers by becoming the „first in the show' are major motives amongst other. Organizations, by deeper understanding of the characteristics of green consumers, may design an efficient and effective sustainable marketing strategy. On the other hand, organization may help alleviating the part of environmental problem by understanding and finally influencing the motives, attitude, behavior and actions of the consumers (Robert & Becon 81).

The consumer buying behaviour is affected by the awareness of the consumer regarding the environment. And this awareness, in turn, affects the decision of the consumers for buying an environmentally friendly product or commonly known as „green products'. However, it has been observed by various researchers that consumers' concern for the environment does not necessarily reflect in its buying decision (Mostafa 447).

It is important for the consumers to be vigilant and particular about its consumption habits. The consumption is called sustainable consumption if it doesn't impact the environment negatively. In other words, the environmental impact of the activities like, purchase, usage and disposal of the product or services, is such that it has either positive impact on the environment or minimal negative impact. This sustainable consumption is closely linked to the social responsibility of an individual and affects the overall decision making regarding the purchase of the product. Now-a-days consumers have become more environmentally conscious and pay heed to the sustainable activities of the companies as well. They take more favourable decisions for the companies having greener image and ethical practices. These consumers are termed as „green consumers'. These consumers support environmental initiatives started by the government and companies and are willing to pay a premium on green products.

To understand the factors responsible for the formation of consumers' attitude towards environmentally friendly products, various researchers have put in their efforts. However, the antecedents of the consumer environmental attitude extracted or identified are still not exhaustive in nature (D'Souza *et al.* 182). Another limitation is the identification of the product or product category for the study on consumer behaviour towards green products is done. There is a huge gap between the research done on the number of products or product category available in the market and the selected product or product category. It has also been observed that the consumers having environmental concern are not ready to buy green products due to the reasons like, perceived effectiveness, high price of the product, not relying on the companies green claim, unavailability of the green product etc. The knowledge of the factors responsible for the development of consumer environmental attitude would enable the marketer to work on their promotional activities and product development area. The knowledge of consumer attitude towards green products will help the marketer influence their perception and attitude in favour of green products which will not only improve the bottom line but would help in sustainable development. To understand the relationship of antecedents of consumer environmental attitude three constructs were selected for the study;(1) Perceived Seriousness of Environmental Problem (PSEP); (2) Eco literacy; and (3) Perceived Consumer Effectiveness (PCE)

3.2 RELEVANT LITERATURE

To study the socio-psychological factors-eco-literacy, perceived consumer effectiveness and perceived seriousness of environmental problem-influencing the environmental attitude of the consumers of Delhi-NCR.The research on consumer environmental attitude is done especially in western world and most of the developing and underdeveloped countries are yet to come up with their concrete research outcomes in the area of consumer and the environment. However, all the researches in the area of consumer and the environment can be divided into four streams. The first research stream deals with macro-marketing issues relating to green consumer attitude and behavioural outcome (Kalafatis et al. 447). It deals with the macro environment (*e.g.* technological, political, and economic) in influencing society's values and beliefs. The macro marketing factors are majorly responsible for the development of attitude and intention to purchase green products. It has been proved that if one adopts the sustainable consumption pattern, it not only improves the situation of deteriorating condition of environment but also improves the quality of life (Kalafatis *et al.* 447). The consumer behaviour is dynamic in nature and most of the consumer categories are changing their behaviour towards a particular product or services very rapidly. As such, their examination is critical for understanding the changes in environmental stability and sustainability (Gardyn 12; Ottman, 1993). While previously the role of consumption was investigated in relation to the environment per se, research in this area attempts to uncover the macro-caveats leading to ecologically-friendly consumption lifestyles. Several studies (*e.g.* Mostafa 448; Roberts& Bacon 568) used the Natural Environmental Paradigm (NEP) scale, originally developed by Dunlap et al. (489), to investigate whether a moregeneral position about society and the environment could have an impact on environmentally conscious consumer behaviour.

Micro-marketing details were discussed in the second stream of the consumers' environmental studies. This includes the characteristics like consumers' eco-literacy, and its perceived seriousness of environmental problems. Here the broad areas of research are to identify the factors responsible for the formation of green consumer attitude and to profile the green consumer. Scholars also focused on categorizing the green consumers based upon demographic and psychological characteristics (Diamantopoulos et al. 468). In this stream, the more emphasis was given to the conceptualizing and operationalizing and measuring the consumer's environmental attitude (Roberts, 1979). Some interesting outcome of this phase was that the demographic attributes play an important role in formation of environmental attitude. As is evident from the study by Roberts (81), the consumers with higher education and higher income show more favourable attitude towards the environment. Other authors like Van Liere and Dunlap (485) and Laroche et al. (509) have concluded that there is an association between the gender type and environmental attitude. Other researches were also been done in this phase taking other factors as inter personal influence, value orientation etc. (Diamantopoulos et al. 473).

The third research stream concentrates on psychological, cultural, societal, andother parameters acting as predictors of consumer ecological concern. This stream got its importance in the era of developing consumerism. The purchase behaviour of people got linked to the social factors and its own value orientation. Subjective and normative functions became more prominent in influencing the behaviour of the general consumer. The role of media cannot be denied in developing the general preference of the consumers. Some of the measures employed were: personal values, such as security, fun/enjoyment, and self-gratification (McCarty & Shrum 455); ethical ideologies, such as deontologism and teleologismsocietal factors, such as liberalism and social altruism (e.g. Straughan & Roberts 565); cultural orientations, such as collectivism and individualism (e.g. Laroche et. al., 198) and personal environmental beliefs, such as faith in others, perceived consumer effectiveness, and susceptibility to normative influence.

Laroche et. al. (515) has worked upon the various socio-psychological characteristics of the consumers which include the education of the consumers, eco-literacy, ecological concern and attitude. It has been observed that the consumers recycling behaviour is also linked with socio-psychological characteristics (McCarty and Shrum, 456). Many authors strive to predict the behaviour towards green products purchase and linked it with the socio-psychological characteristics, ethical ideologies and demographic attributes. Most of the studies conducted in the area of environment marketing established a strong and positive relationship with the above factors however; few authors have concluded negative relationships as well (Banerjee et. al., 183; Laroche et. al., 197; Roberts & Bacon 84).

The environmental attitude of consumers and its associated behaviour have a peculiar relationship which is complex and dynamic in nature. But, to profile a consumer who is environmentally concerned and ecologically conscious it is important to study the relationship of consumer environmental attitude and its associated behaviour (Roberts and Bacon 86).Various researches have proved the positive relationship between the consumer environmentally friendly behaviour and demographic and psychological dimensions (Roberts and Bacon, 85; Stern et al., 283). Other studies have also establish an association between positive environmental attitude and green purchase decision (Kinnear et al.22; Kim and Choi, 597).

Amongst various dimensions, eco-literacy was always taken as a prominent dimension which influence the consumer environmental attitude. Eco-literacy can also be termed as environmental knowledge of the consumers (D'Souza et al. 180). Eco-literacy can be taken in two ways, one the consumers' knowledge regarding the environmental impact of the product while its usage and second the impact on the environment while manufacturing the product. According to Laroche et. al. (201), eco-literacy enjoys a prominent role in influencing the consumer environmental attitude and marketers and environment agencies can use this dimension to influence the environmental attitude and behaviour of the consumer. In India, the knowledge of the environmental degradation and the availability of the environmentally friendly product in the market is still at a very low level. Therefore, environmental knowledge is an important dimension in the hands of marketers and environment agencies to strategies the marketing functions. Thus, wepredict the following;

H₁: Eco-literacy is positively related to consumer environmental attitude.

Perceived consumer effectiveness (PCE): PCE is defined as "the consumer's perception of the extent to which their actions can make a difference in solving environmental problems" (Ellen et al.199). In terms of pro-environmental behaviour, PCE is the extent to which a consumer take actions related to the preservation of the environment for instance, buying a recycled product, carefully handing over the product for recycling after usage, carefully giving due credits to those products which have less environmental impact. In India, there is star rating of almost all electrical appliances, choosing higher stars by a consumer to reduce the energy wastage is the example of PCE. PCE has been included in several studies and is assumed to be an important predictor of ECCB, surpassing all other socio-demographic and psychographic variables (Roberts & Bacon 85; Straughan and Roberts 245). Thus, we may formulate that;

H₂: Perceived Consumer Effectiveness (PCE) is positively related to consumer environmental attitude.

Perceived seriousness of environmental problem (PSEP) is commonly defined as the "individual's awareness of the environmental problems and their willingness to be part of the problem solution" (Dunlap and Jones 488; Chan and Lau, 347). Perceived seriousness as a dimension has been taken by several authors in the name of environmental concern. It has been observed by various authors that people with high level of environmental concern show more favourable environmental attitude and positive green purchase behaviour (Kinnear *et al.* 23; Van Liere and Dunlap487; Roberts and Bacon 83). According to Maloney et al. (11), EC is related to the emotions and knowledge level as well as to a readiness to change behaviour. Kim & Choi (594) point out that consumers that are more concerned about the environment are more willing to purchase green products than those who are less concerned. Thus, we may posit that;

H₃: Perceived Seriousness of environmental problem (PSEP) is positively related to consumer environmental attitude.

Consumers who exhibit the environmentally friendly attitude are of the opinion that the ecological situation on the planet is degrading day by day. They feel that it is the high time that we should do something in order to fix the problem of deteriorating environment (Banerjee & McKeage 149; Schlegelmilch *et al.* 273). However, the cost of exhibiting the environmentally friendly behaviour is very high in terms of paying a premium for these products or compromising with the product performance (*e.g.* cars with lower performance and better fuel efficiency) or buying a product without a comfortable and convenient polythene bag (Kalafatis *et al.* 447; Laroche *et al.* 511). According to Mostafa (466) and Roberts and Bacon (83), the pro-environment behaviour shown by the people is in harmony with the cognitive consistency theory. The theory argues that if an individual is concerned for the deterioration of the environment then his actions will be such that it will minimize the effect of his action on the environment. It has also been shown empirically by Mostafa (477) and Roberts and Bacon (91) that people who are environmentally concerned are more likely to purchase environmentally friendly products and go one step further to influence others to adopt the practice of becoming more environment friendly consumers.

Based on the above, we may propose a conceptual model;

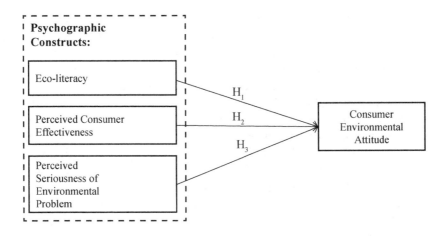

Fig.3.1: Conceptual Framework

3.3 RESEARCH METHODOLOGY

The area of the research is Delhi-NCR which is divided into four zones. Convenience sampling method was used to identify 100 respondents from each zone. However, a total of 347 responses were usable for the purpose of the study. The distribution of the respondents in terms of their demographic profile is given in table no.1.

Table 3.1: Demographic Characteristics of the respondents

Respondents characteristics	Classification	No. of respondents
Age	20-25	112
	26-35	160
	Above 35	75
Educational Qualification	Up to graduation	93
	PG	194
	More than PG	60
Gender	Male	193
	Female	154

A questionnaire was designed as research instrument through which the responses of the people could be recorded and measured. The questionnaire was divided into two sections. First section contained the questions related to three scales of consumer eco-literacy, perceived consumer effectiveness and perceived seriousness of environmental problem. These scales was originally taken from the literature review and then modified accordingly. For instance, perceived consumer effectiveness we used scale developed by Allen *et al.*, (1991) and the scales of eco-literacy and PSEB were developed by (1999) and Dunlop and Jones (2002) respectively. The scale for CEA was developed by Devellis (1995). The reliability of the constructs after finalization of the structured questionnaire is shown in table no.2 and was well in the permissive range *i.e.* Cronebach alpha was more than 0.65 for all the constructs. To ensure the content validity, a pilot study was conducted with first 30 respondents and the suggestions of the respondents and the expert opinion were duly considered incorporated in the final questionnaire. A seven point LikertScale ranging from "strongly disagree" to "strongly agree" has been used to explore the antecedents of the consumers' environmental attitude and their willingness to purchase the green product. Convenience sampling method was employed to collect the primary data from Delhi-NCR region, however, to represent the result fairly, data has been collected from 347 respondents.

Table 3.2: Reliability Measures (Cronbach alpha values) of the constructs

Name of the construct	No of items	Alpha
Eco Literacy (Eco)	7	.821
Perceived Seriousness of Environmental Problem (PSEP)	5	.767
Perceived Consumer Effectiveness (PCE)	4	.713
Consumer Environmental Attitude (AT)	6	.681

3.4 DATA ANALYSIS AND RESULTS

1. *Exploratory Factor Analysis:* Exploratory factor analysis is used to explore the underlying dimensions of the antecedents of the FMCG consumers' environmental attitude. The appropriateness of factor analysis for data has been checked by Barlett's test of sphericity and Kaiser-Meyer-Oklin (KMO) measure of sample adequacy. Barlett's test is applied to test the overall significance of correlation matrix. KMO measure indicates the proportion of variance in the variables, which is a common variance. The results show that the value of KMO statistics is high (.829) and Barlett's test if sphericity is significant (.000) which reveals that data are appropriate for factor analysis. Thus, only the factors having latent roots or Eigen values greater than 1 were considered significant; all the factors with latent roots less than 1 were ignored. Only four components have Eigen values greater than one and total variance accounted for by these factors are 66.36% and remaining variance was explained by other factors Table 3). The factor loadings of all the statements are above .70 confirming the discriminant validity of the constructs.

Table 3.3: Total Variance Explained

Component	Initial Eigenvalues Total	Initial Eigenvalues % of Variance	Initial Eigenvalues Cumulative %	Extraction Sums of Squared Loadings Total	Extraction % of Variance	Extraction Cumulative %	Rotation Sums of Squared Loadings Total	Rotation % of Variance	Rotation Cumulative %
1	3.519	25.997	25.997	3.519	25.997	25.997	3.274	24.881	24.881
2	3.271	19.867	45.864	3.271	19.867	45.864	2.562	22.645	47.526
3	2.542	11.555	57.419	2.542	11.555	57.419	2.454	12.295	59.821
4	1.968	8.945	66.363	1.968	8.945	66.363	2.099	6.542	66.363
5	0.996	7.162	73.525						
6	1.062	5.828	79.353						
7	0.970	2.410	81.764						
8	0.894	2.062	83.826						
9	0.840	1.820	85.646						
10	0.774	1.519	87.165						
11	0.702	1.191	88.356						
12	0.636	1.889	90.245						
13	0.610	1.773	92.018						
14	0.588	1.672	93.690						
15	0.529	1.404	95.094						
16	0.467	1.121	96.216						
17	0.416	0.891	97.106						
18	0.326	0.482	97.589						

Contd...

19	0.294	0.335	97.923					
20	0.246	1.116	99.040					
21	0.147	0.670	99.710					
22	0.064	0.290	100.000					
Extraction Method: Principal Component Analysis.									

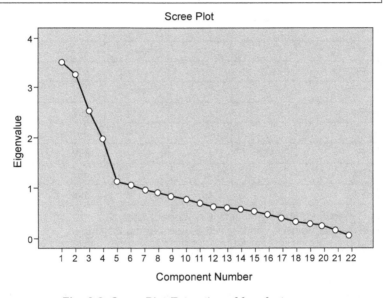

Fig. 3.2: Scree Plot-Extraction of four factors

Table 3.4: Rotated Component Matrix[a]

	Component			
	1	2	3	4
EA3	.895			
EA5	.833		0.412	
EA4	.832			
EA2	.766			0.529
EA6	.690			
EA1	.449			

Contd...

ECO7		.750	
ECO6		.732	0.422
ECO2		.712	
ECO5		.667	
ECO1		.543	
ECO4		.469	
ECO3		.696	
PC1			.869
PC4			.839
PC3			.751
PC2			.618
P4			.697
P2			.641
P5	0.467		.632
P3			.632
P1			.598
Extraction Method: Principal Component Analysis. Rotation Method: Varimax with Kaiser Normalization.			
a. Rotation converged in 6 iterations.			

The varimax has been used that rise to maximize the variance of each of the factors so that the total amount of variance account for is redistributed over the three extracted factors. The Rotated Component Matrix reveals four factors which represent the four broad dimensions of consumers' environmental attitude termed asEco-literacy(Eco), Perceived Seriousness of Environmental Problem (PSEP), Perceived Consumer Effectiveness (PCE),Consumer Environmental Attitude (CEA)derived from 22 statements. To confirm the three dimensional antecedents of consumers' environmental attitude (as obtained by EFA), a confirmatory factor analysis (CFA) is done.

2. *Confirmatory factor analysis (CFA):* In this paper, confirmatory factor analysis is run through IBM-AMOS software. Confirmatory factor analysis (CFA) allows the researcher to test the hypothesis that the relationship between observed variable and underlying latest construct exists. Confirmatory factor analysis (CFA) is a measurement model that estimates latent variables based on observed indicator variables and also checks the reliability of the model. CFA is the technique to find out the exact *relationship between* the common *factors* and the *items* used to measure *them as well as* the *linkages* among the *factor with reliability* (Salisbury et. al., 2001;Joreskog and Sorbom, 1989). The table 5 represents the convergent and discriminant validity measures of four extracted constructs taken together in CFA. As shown in the results composite reliability (CR) is more than 0.7 as well as greater than average variance extracted (AVE). This ensures the existence of convergent validity in the instrument. In addition to this, average variance extracted of each construct is greater than MSV and ASV statistics for the constructs which ensures the existence of discriminant validity.

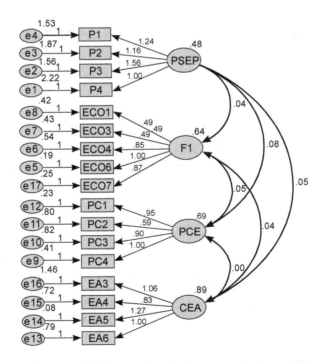

Fig. 3.3: CFA for the four factors extracted from EFA

Table 3.5: Convergent and Discriminant Validity measures of four constructs

	CR	AVE	MSV	ASV	PCE	PSEP	ECO	CEA
PCE	0.791	0.597	0.018	0.008	0.705			
PSEP	0.811	0.673	0.018	0.008	0.133	0.523		
ECO	0.821	0.509	0.007	0.005	0.080	-0.085	0.640	
CEA	0.792	0.760	0.001	0.000	0.006	0.001	-0.031	0.678

For the model to be considered fit it should support the criteria of the various indices such as $X^2/df \leq 3$, Goodness of Fit (GFI) ≥ 0.90 (Joreskog and Sorbom 1989), Adjusted Goodness of Fit (AGFI) ≥ 0.80, Comparative Fit Index (CFI) ≥ 0.89 and RMSEA≤ 0.10. There is no single measure to conclude about the validity of the model.

Table 3.6: Goodness of Fit Indices

Model	NPAR	CMIN	DF	P	CMIN/DF
Default model	50	725.716	303	0	2.39510231
Saturated model	253	0	0		
Independence model	22	3064.231	231	0	13.265

Table 3.7: Goodness of Fit Indices and Adjusted goodness of Fit indices

Model	RMR	GFI	AGFI	PGFI
Default model	0.087	0.911	0.887	0.691
Saturated model	0	1		
Independence model	0.28	0.531	0.486	0.485

Table 3.8: Root mean square error of approximation

Model	RMSEA	LO 90	HI 90	PCLOSE
Default model	0.094	0.088	0.101	0
Independence model	0.188	0.182	0.194	0

Table 3.9: Comparative Fit Indices

Model	NFI	RFI	IFI	TLI	CFI
	Delta1	rho1	Delta2	rho2	
Default model	0.731	0.693	0.782	0.75	0.88
Saturated model	1		1		1
Independence model	0	0	0	0	0

The above Table nos.,6, 7, 8 and 9 indicate that the statistical indices required for a model to be fit and can be considered for further study are well in the required range.

3.5 DISCUSSION, LIMITATIONS & OUTCOMES

The study contributes to the marketing research literature in multiple ways. First, it is one of the very few studies done in India and abroad wherein five antecedents of consumer environmental attitude have been taken together and their relationship with consumer environmental attitude is checked and established using confirmatory factor analysis. These three antecedents, Eco-literacy(Eco), Perceived Seriousness of Environmental Problem (PSEP), Perceived Consumer Effectiveness (PCE), have shown a positive relationship with consumer environmental attitude. Second, this study simultaneously examines the societal, cultural and ethical issues in forming environmental attitude by consumers. Third, this study draws a clear cut distinction between environmental attitude and environmental behavior in terms of willingness to purchase environmentally friendly products whereas most of the previous research have used both the terms interchangeably. Finally, this study combines micro and macro green marketing issues, often examined separately by previous research.

The main findings of the study can majorly be divided into two sections. First, related to the relationship of three antecedent, Eco-literacy(Eco), Perceived Seriousness of Environmental Problem (PSEP), Perceived Consumer Effectiveness (PCE) with consumer environmental attitude (CEA) construct. The result is consistent with the findings of the various authors in previous researches. Laroche et al. (516) argued that individual eco-literacy (environmental knowledge) plays an important role in shaping the attitude and finally influencing the green purchase behaviour. In addition to that, Bandura's (233) social cognitive theory supports the findings, which argues that there is, triadic relationship between personal, environmental and behavior factors'. The findings show that the level of eco-literacy (that is the environmental knowledge) significantly correlates with the consumer attitude and behavior tendencies (Laroche et. al. 517; Roberts and Bacon84). The findings furthers reinforces that if the consumer has sufficient knowledge about the environment and the causes of environmental pollution, then their overall awareness level towards green products will increase and thus, they will form more favourable attitude towards environmentally friendly products. Therefore, it is crucial for marketer and public relation practitioners to disseminate information regarding the environmental impact of their product and educate consumers through communication initiatives such as environmental support campaigns. However, environmental knowledge does not necessarily mean that the consumer will make a green purchase.

With regard to consumers' susceptibility to the interpersonal influence and environmental attitude, the findings of the study coincide with the previous studies done by Beardone et.al., (198) and Stafford and Cocanaugher (234). This means that interpersonal influence (IPI) from peers and family members will have a significant effect on the consumer's attitude towards the environment. However, this result deviates from the findings of the study done by Isaac Cheah and Ian Phau (466). Talking about the value orientation (VO), the study shows that collectivism plays a positive role in framing the positive attitude towards the environment. In other words, collectivist consumers tend to exhibit more friendly behavior towards the environment than the individualistic consumers. These results correspond to the findings of McCarty and Shrum (214) and Cheah and Phau (467) wherein it was proposed that collectivist consumers tend to be friendlier towards the environment than the individualistic consumers. Sarigollu (133) proposed that long-term oriented people have a favorable attitude towards the protection of natural environment. Findings of our study coincide with the literature. Long-term oriented people exhibit favorable attitude towards the natural environment. The overall impact of increasing consumption on the environment has been witnessed and recorded by the regulatory bodies.

As for the association between population size and the climate change, the relationship is not simple. However, a simple thumb rule suggests that with the increase in the population, the earth natural resources have to be used in greater quantity. Be it arable land, residential purpose land, industrial land potable water or any other resource, the usage will increase the carbon emission and hence the global warming problem has to be addressed. Every year the requirements of potable water, arable land and land for other purposes are increasing many folds. Between 1990 to 1995, the water consumption rose six folds which is more than double the rate of growth.

The distribution of population across the globe also affects the climate change positively. Since there is more fertile land in developing nations and less fertile land in developed nations, the majority of the population has shifted to developing nations. More and more people are migrating from less developed areas to more developed regions. Every year around 2to 4 million people migrate internationally. Within the country people are migrating from rural area to urban area in search of work. New data on the demography of people suggests that globally, the population is urbanized. The researchers and social scientists have forecasted that the trend of increase in population and accumulation of population in urban areas will not stop in the near future. More and more people will be shifting which means that more houses, more water and land, more resources are required in order to maintain the social balance. All these exercise will definitely put pressure on the environment and if strong measures are not taken by the authorities in terms of regulation and control, it will lead to alarming situation of environmental pollution.

The composition of the population also plays an important role in affecting the ecological system of any country or the continent per se. For instance, India has major percentage of young people (under the age group of 30). It has been observed that young people have more propensity to migrate from rural to urban area than the older people. So, India now a days experiencing an increased level of migration from rural areas to urban areas. Not only that, few cities have become the magnets for the young generation and these magnet cities have also become the prime reason for the migration. Therefore, more migration and concentration of people at a particular area calls for better management of resources which raises the alarm for addressing the environmental concern of that area.

Another important dimension of global population which affects the environment is ‚income' or the economic development'. Across countries, the relationship between economic development and environmental pressure resembles an inverted U-shaped curve. It has been observed that the economies with developed status exert more pressure on the environment by emitting more gases in the environment due to their industrialization and high consumption pattern, whereas the economies with relatively lower development status exert less pressure on the environment. There is greater need to address the issue of growing environment pollution in the developed and developing nations. Innovative methods have to be adopted using new technology and energy conservation.

There is classical role of cultural role of cultural dimension of the population which affect the environment positively. It includes the value system of the population and socio-psychological dimensions as well. For instance, if cultural attitude of the people is positive towards the wildlife or forest conservation, they will do their best to protect the natural environment. Their actions will be oriented towards the preservation of the natural resources and keeping the value system of the family and the community intact. Even, most of the government initiatives' success depends upon the support of general public. Marketers and other government agencies have a herculean task to bring the cleanliness and developing a habit of thinking of the environment before taking any action (even for a purchase decision for a product or service).

The policy implications of the outcome of various researches done in the field of consumer attitude towards the environment and intention to buy green product is very controversial in nature. However, the macro marketing factors play vital role in deciding and influencing the consumer environmental attitude, but the evidence for it is very superficial. In case of micro-marketing factors, number of studies have suggested very concrete results but specific to particular country or region. Therefore, the general public has to be made sensitive about the environmental impact of their day to day activities and consumption pattern. In order to accomplish this goal, eco literacy is the most important factor. Eco-literacy can be understood in two basic levels, first, the knowledge of the general consumer regarding the environmental impact of the product while manufacturing and procuring raw-material for the same. And second, the knowledge of environmental impact of the product while using and disposing the product after its usage.

The study has various limitations which may be considered as areas of further research on the same issue. Convenience sampling technique is one of the limitations as this might not truly represents the entire population of Delhi-NCR in India. Socio-demographic statistics can be seen to fluctuate and hence the associate consumer attitude towards green products may deviate from the findings of the model over a period of time. Volsky et. al. (129), proposed that people belonging to age group of 45-54 can be identified as ‚ultra-green' and these people are the driving force of environmentalism. Whereas D'souza (181), segmented the consumers in two groups emerging green (age 25-39) and potential future market participants (age 18-24). Antecedents of environmental attitude in this study are not exhaustive. Other antecedents which are associated with attitude formation are subjective norms, past experience, perceived seriousness of environmental problem. Cultural, political and ethical parameters can also be taken for future research. There are factors which may moderate the relationship between environmental attitude and willingness to buy green products. Product necessity, product involvement and pricing concepts are few of moderating factors which may influence the attitudinal tendencies.

REFERENCES

1. Akehurst, Gary., Afonso,Carolina., Gonçalves,Helena M. "Re-examining green purchase behaviour and the green consumer profile: new evidences". Management Decisions 50. 5. (2012): 972-988. Print.
2. Atkin, Leanne. "Environmnetal Attitude and the green consmer profile: New Evidences." Management Decisions (2006): 972-988. Print.

3. Ang, Swee H., Cheng, Peng S., Lim, Eli A.C. and Tambyah, Siok K. "Spot the difference: consumer responses towards counterfeits", Journal of Consumer Marketing 18.3 (2001): 256-278. Print
4. Bandura, Albert. Social Foundations of Thought and Action: A Social Cognitive Theory. Newsealand: Prentice-Hall, Englewood Cliffs, 1986. Print.
5. Banerjee, Bobby S. Corporate environmentalism: The construct and its measurement. Journal of Business Research 55 (2002): 177–191. Print
6. Banerjee, Bobby S. and Mckeage, Kim. How green is my value: Exploring the relationship between environmentalism and materialism. Advances in Consumer Research 21. (1994): 147–152. Print.
7. Brown, David. "It is good to be green: environmentally friendly credentials are influencing business outsourcing decisions", Strategic Outsourcing: An International Journal 1.1. (2008): 87-95. Print
8. Chan, Kara. "Market segmentation of green consumers in Hong Kong", Journal of International Consumer Marketing 12. 2. (1999): 7-24. Print.
9. Chan, Ricky, Lau, Lloyd. "Antecedents of green purchases: a survey in China", Journal of Consumer Marketing 17. 4 (2000): 338-57. Print
10. Cheah, Isaac., Phau, Ian. "Attitudes towards environmentally friendly products: The influence of ecoliteracy, interpersonal influence and value orientation", Marketing Intelligence & Planning 29. 5 (2011): 452 – 472. Print.
11. Crane, Andrew. "Facing the backlash, green market and strategic reorientation in the 1990s", Journal of Strategic Marketing 8. 3 (2000): 277-96. Print.
12. Crane, Andrew. "Marketing and the natural environment: what role for morality?", Journal of Macromarketing 20. 2(2000): 144-54. Print.
13. D'Souza, Clare. "Ecolabels programmes: a stakeholder (consumer) perspective", Corporate Communication: An International Journal 9. 3 (2004): 179-88. Print.
14. DeVellis, Robert F. Scale Development, Thousand Oaks, CA: Sage Publications, (1991). Print.
15. Diamantopoulos, Adamantios, Schlegelmilch, Bodo B., Sinkovics, Rose R. Can socio-demographics still play a role in profiling green consumers? A review of the evidence and an empirical investigation. Journal of Business Research 56 (2003): 465–80. Print.
16. Dunlap, Riley E. and Jones, Ride E. "Environmental concern: conceptual and measurement issues". in Dunlap, Riley E. and Michelson, Willium. Ed. Handbook of Environmental Sociology,Westport, CT, Greenwood Press, 2002, 482-524. Print
17. Ellen, Palm S., Wiener, Joshua L. and Cobb-Walgren, C. "The role of perceived consumer effectiveness in motivating environmentally conscious behaviors", Journal of Public Policy & Marketing10. Fall.(1991): 102-117. Print.

18. Gardyn, Rendevil. "Eco-friend or foe?" American Demographics 25. 8. (2003): 12. Print.
19. Hofstede, Geert. Culture's consequences: International differences in work-related values. Beverly Hills, CA: Sage.1980. Print.
20. Kalafatis, Stavrose P., Pollard, Michael., East, Robert. and Tsogas, Marcos H. "Green marketing and Ajzen's theory of planned behaviour: a cross-market examination".Journal of Consumer Marketing 16. 5(1999): 441-460. Print.
21. Kilbourne, William. and Pickett, Gregory. How materialism affects environmental beliefs, concerns, and environmentally responsible behaviour. Journal of Business Research 61.(2008): 885–893. Print.
22. Kim, Young. and Choi, Soburt. "Antecedents of green purchase behavior: an examination of collectivism, environmental concern, and perceived consumer effectiveness". Advances in Consumer Research 32.(2005): 592-599. Print.
23. Kinnear, Thomas C., Taylor, James R. and Ahmed, Sadrudin A. "Ecologically concerned consumers: who are they?". Journal of Marketing 38.(1974): 20-24. Print.
24. Laroche, M., Bergeron, J. and Forleo, G.B. (2001), "Targeting consumers who are willing to pay more for environmentally friendly products". Journal of Consumer Marketing, Vol. 18 No. 6, pp. 503-520. Print.
25. Laroche, Michel., Bergeron, Jasmin. and Forleo, Guido B. "Targeting consumers who are willing to pay more for environmentally friendly products". Journal of Consumer Marketing 18. 6. (2001): 503-20. Print.
26. Laroche, Michel., Toffoli, Roy., Chankon, Kim. and Muller, Thomas E. "The influence of culture on pro-environmental knowledge, attitudes, and behavior: a Canadian perspective".Advances in Consumer Research. 23.(1996): 196-202. Print.
27. Leonidou, Leonidas C., Leonidou, Constantinos, N. Kvasova, Olva. "Antecedents and outcomes of consumer environmentally friendly attitudes and behavior".Journal of Marketing Management, 26. 13-14(2010): 1319–1344. Print.
28. Manaktola, Kamal. and Jauhari, Vinie. "Exploring consumer attitude and behavior towards green practices in the lodging industry in India".International Journal of Contemporary Hospitality Management 19. 5(2007): 364-77. Print.
29. Menon, Ajay., & Menon, Anil. Enviropreunerial marketing strategy: The emergence of corporate environmentalism as marketing strategy. Journal of Marketing. 61. (1997): 51–67. Print.
30. Mostafa, Mohamed.M. A hierarchical analysis of the green consciousness of the Egyptian consumer. Psychology and Marketing. 24. 5 (2007): 445–473. Print.
31. Ottman, Jacquelyn C. Green Marketing: Challenges and Opportunities for the New Marketing Age. McGraw-Hill, New York, NY.1993. Print.

32. Polonsky, Michael J. and Rosenberger, Philip J. III. "Reevaluating green marketing: a strategic approach".Business Horizons 44. 5. September/October. (2001): 21-30. Print.
33. Roberts, James A. and Bacon, Donald R. "Exploring the subtle relationship between environmental concern and ecologically concerned consumer behavior".Journal of Business Research 40 (1997): 79-89. Print.
34. Straughan, Robert D. and Roberts, James A. "Environmental segmentation alternatives: a look at green consumer behaviour in the new millennium", Journal of Consumer Marketing 16. 6 (1999): 558-575. Print.
35. Volsky, Richard P., Ozanne, Lucie K. and Fontenot, Renee J. "A conceptual model of US consumer willingness-to-pay for environmentally certified wood products", Journal of Consumer Marketing16. 2.(1999): 122-140. Print.

Chapter 4
Water Policy in India: Building Blocks for Synergy with Science, Technology and Innovation

Venkatesh Dutta, Karunesh Kumar Shukla, Subhash Chander

Abstract: *India's mainstream water policy reform has largely focused on institutions, governance and resource development. The importance of technology and innovation in water resource management has received less attention. The type of innovation and the changes which must be made within the current set-up has not been debated previously and the conditions that enable or hinder those changes remain unclear. India is facing many water challenges due to various factors such as increasing population, urbanization, changing lifestyle, changing frequencies of rainfall events, excess withdrawal of groundwater, non-scientific water allocation, quality deterioration, wastage and inefficient use. The role that science, technology and innovation (STI) can play in delivering water policy objectives for better decision making for water resource management, and operational pathways to address these challenges provides a highly significant lens for approaching assessment of broader water policy framework. India's water policy has been drafted keeping in view certain basic principles such as integrated perspective on society, economy and the environment; equity and social justice in allocation, good governance through informed decision-making, community management under public trust, ecological needs, basin as the basic hydrological unit for planning, demand management and consideration of the local, geo-climatic and hydrological situation. Implicit in all these principles is the need for integrating evidence-based science into policy design and delivery. Role of better STI integration within the water policy framework is critical for operationalizing these principles in a* *holistic manner. This paper argues that integrated and effective water management cannot be achieved if information, technology, people and ecosystems are conceptualized as separate entities. There are still some basic issues that are tentative about how to put principles of integrated water resources management into practice. The paper delineates various transformative changes that are required in the current management system that governs the water resource development and use as outlined in the National Water Policy of 2012. The paper also identifies key policy areas, discusses challenges and opportunities for strengthening the water policy framework taking into account instances of synergy between the NWP 2012 and the STI Policy 2013.*

Keywords: *Mainstream Water Policy, Water Management System, Frequencies of Rainfall Events, Equity & Social Justice, Science, Technology & Innovation Policy*

4.1 INTRODUCTION

Sustainable and equitable use of water has been historically advocated in India by cultural adaptation to water availability through water conservation means, agricultural systems and cropping patterns. Although significant progress is being made in ensuring sufficient availability of water for society and ecology, various challenging issues continue despite a number of institutions established and programs initiated for sustainability of water resources. Due to large population, industrial needs and urbanization and increasing societal demand, India continues to struggle to meet its water requirements. Though India possesses only four percent of the global fresh water resources, it has to provide drinking water for 16 percent of the world population. Further, water security is under intense pressure in urban areas and metropolitan cities, and the very nature of urbanization adds to water stress situations both from a quantity and quality perspective. The problems will become more intense due to overuse and pollution thereby disturbing the quality and the natural cleansing capacity of water resources (Biswas and Tortajada, 2010). Some of the institutions involved in the governance of water resources in India have been formed by the deliberate „designing' and largely followed business-as-usual approaches without emphasizing on the ability of these institutions to contribute to the demand or supply management challenges (Saravanan, 2015). There is also a policy vacuum – how to translate the scientific knowledge coming out from these institutions into specific policy actions (Katyaini and Barua, 2015). Policies have been formulated in a way that does not reflect the supply and demand management challenges and fail to provide adequate and appropriate solutions. The existing regulatory framework has partially succeeded in „protection' of the water resource from over-allocation and has supported mostly „infrastructure' and „supply-side' based solutions in mitigating water scarcity (Kulkarni *et.al.*, 2015).Researchers have been trying to balance sometimes conflicting environmental, social, and economic goals in water management and governance (Saravanan, McDonald and Mollinga, 2009). The current focus of water policy discourse falls majorly on integrated water resources management, ecosystem-based approaches, and adaptive management. These principles were considered while drafting India's National Water Policy of 2012 (NWP 2012) and the policy further prompted a new way of perceiving and acting with water in India. Though the NWP 2012 is well drafted, the integration between water, energy, climate and agricultural policy objectives is generally limited. The STIP 2013 can play a positive role in several ways in translating scientific knowledge into policy actions and help mitigate water scarcity.For this to happen the integration should be strongly complementing or even replacing established institutional governance concepts.

Though the NWP 2012 outlines the various principles of sound management, the current approach of dealing with water largely follows a „command-and-control paradigm', where the goal of water management is to maximize resource exploitation by reducing natural variability. According to Schoeman *et.al.* (2014) this approach is "typified by centralized, sectoral institutions, limited stakeholder involvement and expert led problem solving focused on technical engineering solutions". Water policy innovation based on technology is largely recognized as needed to transform socio-economic and political circumstances, yet very little understanding exists about the type or form that these policy innovations should take across the water policy literature (Moorea *et.al.*, 2014).There is also a lack of understanding of how policies change in the absence of a well-established governance system and the subsequent effect on practice. It would be very useful to learn how better integration of STI variables can contribute to the success of water policy and hence allow to significantly improve the current practices and processes governing water allocation and use. Section 15.1 of the NWP 2012 mentions that "research and advancement in technology shall be promoted to address issues in water sector in a scientific manner – innovations in water resources sector should be encouraged, recognized and awarded". The 12th Five-Year Plan (2012–2017) has emphasized on water mapping, watershed development, and has involved NGOs, and tracked loopholes where irrigation capacity can be increased. Since water is a State subject in the Federal Constitution, State Governments are expected to play a large role in these efforts.

4.2 WATER CHALLENGES IN INDIA

India faces the problems of low irrigation water efficiency and under-utilization of irrigation water potentials. During the low flow regimes, increased irrigation is further reducing water availability. Until recently the irrigation development focused mainly on large numbers of publicly funded mega storage projects with nominal attention on demand management. Considering per capita water availability as an indicator, „water stress' is already beginning to show in India and given the projected increase in population by the year 2025, the annual per capita availability is likely to drop to the level of water scarcity (Sampath *et.al.* 2008). The average annual per capita water availability in India is continuously decreasing and at present we are in water-stressed condition; however, about 15.2% of total geographical area (49.81 Million hectare) of the country is occurring in flood prone areas.

India has a reasonably good endowment of water resources with an average rainfall of 1083 mm equivalent to 3560 billion cubic metre (BCM)/year (INAE, 2012). However, the available amount is substantially less and is estimated at 1869 BCM in the form of surface water and groundwater. In India, about 85% of rural, 50 % of urban and industrial and about 55% of irrigation water demand is met from groundwater resources only (Najeeb, 2013). According to the CGWB report (2009), the groundwater draft in the country is 243 BCM as against the net annual groundwater availability of 396

BCM, and the annual replenishable groundwater resource for the entire country as evaluated in the CGWB report (2012-13) is 431 BCM. About 57% of the annual recharge of groundwater is contributed by the monsoonal rainfall only as it is the major source of groundwater recharge in the Indian subcontinent; the overall contribution of rainfall is 68%, rest of 32% is met from the other resources (e.g. ponds, canals, tanks etc.). Participatory irrigation management in India is regarded still as experimentation in diverse socio-economic settings with mixed results (Poddar et al., 2014). It is widely argued that innovative and emerging technologies for water storage, treatment and irrigation practices are essential to promote inclusive growth.

4.2.1 Water availability and scarcity

Official estimates of the Ministry of Water Resources (MoWR) have put total utilizable water at 1123 BCM as against the current use of 634 BCM, reflecting a surplus scenario (Planning Commission, 2010). Narsimhan (2008) calculated the water budget using an evapotranspiration rate of 65 per cent as against the 40 per cent used in official estimates.

Table 4.1: India's water budget (Source: Narsimhan, 2008)

	Analysis based on MoWR	Analysis based on MoWR Estimates based on worldwide comparison
	(Values in BCM)	
Annual rainfall	3,840	3,840
Evapotranspiration	3,840-(1,869+432) =1539 (40%)	2,500 (65%) World-wide comparison
Surface runoff	1,869 (48.7%)	Not used in estimate
Groundwater recharge	432 (11.3%)	Not used in estimate
Available water	2,301 (60%)	1,340 (35%)
Utilizable water	1,123 (48.8% of 2,301)	654 (48.8% of 1,340)
Current water use	634	634

The above Table 1 gives the comparative picture of the water budget as per the two estimates. It is pertinent to note that there exists a considerable temporal and spatial variation within the country with respect to water availability. The per capita average annual water availability in the country is reducing progressively due to increase in population. Since independence, India has made significant progress in increasing the storage potential of the available water by building dams on various rivers. A total storage capacity of 212.78 BCM has been created in the country through major and medium projects.

Table 4.2: Average annual per capita water availability
(Source: UNICEF 2013)

Year	Population (Million)	Per capita Average Annual Availability (m³/year)
2001	1029 (2001 census)	1816
2011	1210 (2011 census)	1545
2025	1394 (Projected)	1340
2050	1640 (Projected)	1140

The per capita water availability figures given above are the national average figures while the position is quite different in the individual river basins. A per capita availability of less than 1700 cubic meters (m³) is termed as a water-stressed condition while per capita availability below 1000 m³ is termed as a water scarcity condition. In India, the an average annual per capita availability trend of water is continuously decreasing from 2001 and proposed to follow this trend till 2050 (Fig.1).

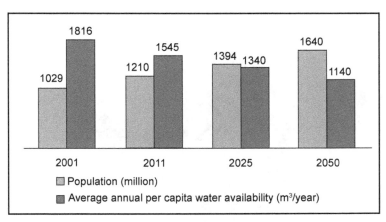

Fig. 4.1: Average annual per capita availability trend of water in India

Rainfall and snowfall are the ultimate sources of water for meeting needs of drinking, irrigation, ground water recharging, rainfed agriculture, and environmental flows, flood and farm income securities. Water is the essential component of life. Unless it is in balanced quantity, any deficit or excess may cause physiographic imbalance. In India, droughts and floods are the frequent natural calamity which finds in all the great epics of the country (Fig.2 & Fig. 3).

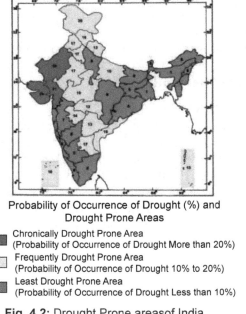

Probability of Occurrence of Drought (%) and Drought Prone Areas

■ Chronically Drought Prone Area
(Probability of Occurrence of Drought More than 20%)
□ Frequently Drought Prone Area
(Probability of Occurrence of Drought 10% to 20%)
■ Least Drought Prone Area
(Probability of Occurrence of Drought Less than 10%)

Fig. 4.2: Drought Prone areas of India

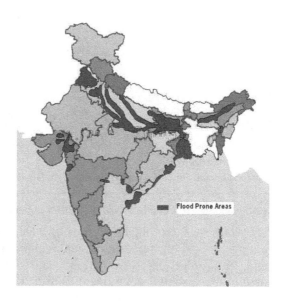

Fig. 4.3: Flood Prone areas of India

(Source: Hydro-met Division, IMD)

As per report of National Disaster Management Authority (NDMA) 2012, in India about 49.81 Million hectare (15.2% of total geographical area) is flood prone and on an average 10-12 Million hectare is actually affected every year causing a range of miseries.

4.2.2 Water Demand:

According to the Census of India 2001, there were 498 Class I and 410 Class II cities in the country; the estimated total population of these cities in 2008 was 257.75 million and total water supply was 48090.88 MLD. Average per capita water supply in Class I and Class II cities was 179.02 and 120.79 liter per day (CPCB, 2009a). Table 3 gives the water demand estimate of diverse sectors by two different agencies- the Standing Sub-committee of the Ministry of Water Resources (MoWR) and the National Commission on Integrated Water Resources Development (NCIWRD).

Table 4.3. Projected Water Demand (in BCM) for Various Sectors

Sector	Standing Sub-committee Report of MoWR			NCIWRD		
Year	2010	2025	2050	2010	2025	2050
Irrigation	688	910	1072	557	611	807
Drinking water	56	73	102	43	62	111
Industry	12	23	63	37	67	81
Energy	5	15	130	19	33	70
Others	52	72	80	54	70	111
Total	813	1093	1447	710	843	1180

4.2.3 Urban Water & Sanitation:

Water supply is not uniform across all states or cities of India. As per census 2011, 70.6 percent of urban households use tap water for drinking, of which 62 percent is treated and 8.6 percent remains untreated. 20.8 percent use water from hand pump / tube well, 6.2 percent use water from well. The census 2011 data shows that 18.6 percent households in urban India do not have latrines within the premises. 6 percent urban households use public latrines. It is reported that about 78 per cent of India's urban population has access to safe drinking water (CPCB, 2009a). However, safe drinking water, sanitation, and a clean environment are essential needs and basic human rights, and India's policy framework very well recognizes this. The provision of adequate drinking water for the entire urban and rural populations is one of the priorities of the NWP 2012. Provision of clean drinking water, sanitation and a clean environment are vital to improve the health of our people and to reduce incidence of diseases and deaths (Planning Commission, 2008).

4.2.4 Water Pollution:

According to the WHO estimates, over 97 million Indians lack access to safe water, second only to China. As a result, the World Bank estimates that 21 per cent of communicable diseases in India are related to unsafe water. The most polluting source for rivers is city sewage and industrial waste discharge. Presently, only about 10 per cent of the wastewater generated is treated; the rest is discharged as is into our water bodies. Due to this, pollutants enter rivers, lakes and the groundwater. Such water, which ultimately ends up in our households, is often highly contaminated and carries disease-causing microbes. Agricultural runoff, or the water from the fields that drains into rivers, is another major water pollutant as it contains fertilizers and pesticides. Water pollution has not been adequately addressed in any policy in India, either at the central or the state level. In the absence of a specific water pollution policy, which would also incorporate prevention of pollution, treatment of polluted water and ecological restoration of polluted water bodies, government efforts in these areas would not get the required emphasis and thrust.

The CPCB's analysis of water quality monitoring results from 1995 to 2009 indicates microbial contamination as the predominant form of pollution in surface water bodies in India (CPCB, 2009b). It was observed that nearly 64 per cent of the surface observations had biochemical oxygen demand (BOD) less than 3 mg/l, 19 per cent between 3-6 mg/l and 17 per cent above 6 mg/l. There is legislation in India that addresses the prevention and control of pollution-for example, the Water Prevention and Control of Pollution Act of 1974, Water Act of 1977 (amended in 1988 as the Water Prevention and Control of Pollution Act), and the umbrella legislation, the Environment (Protection) Act or EPA (1986). Although the Indian government is working more proactively on the increasing threat of water pollution, it will take far more than political will for these actions to translate into concrete measures resulting in improved water quality.

4.3 NATIONAL WATER POLICY 2012

India's comprehensive National Water Policy (NWP), was formulated in 1987, and was revised in 2002 and updated in 2012, which is the current National Water Policy for India. Along with the finalization of NWP 2012, substantial evidence to support principles of integrated water resource management also came from the UN status report on Integrated Approaches to Water Resources Management (UN, 2012). The objective of the NWP is to take cognizance of the existing situation, to propose a framework for creation of a system of laws and institutions and for a plan of action

with a unified national perspective. Some key principles of NWP 2012 are outlined below:
1. Equitable and sustainable development of water
    - To provide drinking water and sanitation for all
    - For the sustenance of ecosystems
    - To achieve food security
2. Maintenance of integrity of all elements of the water cycle such as-
    - Precipitation
    - Evapotranspiration
    - Runoff
    - Soil moisture
    - Ground water
3. Water needs to be managed
4. To promote its conservation
5. For efficient use
6. To reduce pollution
    - Basin as a basic unit for planning of water, to assess and evaluate -
    - Impact of climate change
    - Deterioration of water quality
7. Good governance through informed decision making and practice of-
8. Water audit and accounts for urban, rural and other use of water

For the implementation of these principles, National Water Board has been envisaged to prepare a plan of action based on the NWP 2012, as approved by the National Water Resources Council, and to regularly monitor its implementation. The State Water Policies have also been drafted by several states in accordance with this policy keeping in mind the basic concerns and principles as also a unified national perspective. To tackle the various water related challenges and implement the NWP 2012, there is a need to evolve a National Framework Law governing the exercise of legislative and/or executive powers by the Centre, the States and the local governing bodies. The NPW 2012 mentions that, the water framework law "should lead the way for essential legislation on water governance in every State of the Union and devolution of necessary authority to the lower tiers of the government to deal with the local water situation".

4.4 SYNERGY BETWEEN STI POLICY 2013 AND NATIONAL WATER POLICY 2012

In the assessment of policy landscape that govern the management of water in India, it is argued that, identifying cross-sectoral, multi-scale policy interdependencies, such as between NWP 2012 and STIP 2013 can reduce mismatches in policy making, increase synergies, and therefore promote water resources security. Various researchers have previously discussed this approach where water security issues are seen inextricably linked with other social, economic and political issues such as agriculture, climate change, energy production and human livelihoods (Benson *et al.*, 2015; Bizikova *et.al.*, 2013; WEF, 2011).

India's Scientific Policy Resolution of 1958, the Technology Policy Statement of 1983, the Science and Technology Policy of 2003 and the Science, Technology and Innovation Policy of 2013 categorically state, "Science and technology have an unprecedented impact on economic growth and social development". They affirm that one of the key aims of these policies is to secure for the people all the benefits that can accrue from the acquisition and application of scientific knowledge. The objective is to ensure the security of the people in terms of food, agriculture, water, nutrition, the environment, health and energy on a sustainable basis, with special emphasis on „equity' in development, so that the benefits of technological growth reach the majority of the population, particularly the disadvantaged, leading to an improved „quality-of-life' for every citizen (DST, 2013). According to the STIP 2013, the guiding vision of aspiring Indian science, technology and innovation enterprise is "to accelerate the pace of the discovery and delivery of science-led solutions for faster, sustainable and inclusive growth".„Innovation for inclusive growth implies ensuring access, availability and affordability of solutions to as large a population as possible.' It says that the policy will drive both investment in science and the investment of science-led technology and innovation in select areas of socioeconomic importance. The STIP 2013 in principle intends to position India among the top five global scientific powers by 2020, assist S&T-based high-risk innovations through new mechanisms; facilitate partnerships among stakeholders for scaling successes of R&D; and trigger changes in the mindset and value systems to recognize, respect and reward performances which create wealth from S&T derived knowledge.

STIP 2013 recognizes synergy between innovation and S&T and it is important that all dimensions of STI could be vertically integrated into social, economic, ecological and cultural values. Since the policy focuses on both "people for science and science for people", it assumes far greater importance in the wake of failure of past conventional approaches in achieving effective water management. Innovation led S&T assumes center stage in the overall water management in creating a ‚new water paradigm' that integrates technology and society with integration of sectors, issues and disciplines. India's Science and Technology Policy of 2003 called for ‚integrating programmes of socio-economic sectors with the national R&D system', while articulating upon the need for technological innovation and creation of a national innovation system. This is also the intention of the Water Policy 2012 through creation of a system of laws and institutions and for a plan of action with a unified national perspective.

Table 4.4: Components of STIP 2013 and areas where water policy directives can be synergized

	Components of STIP 2013				
R&D and innovation landscape	Facilitating private sector investment in R&D centres in India and overseas	Permitting multi stakeholders' participation in the Indian R&D system	Treating R&D in the private sector at par with public institutions for availing public funds	Bench marking of R&D funding mechanisms and patterns globally	Aligning Venture Capital and Inclusion Innovation Fund systems

Contd...

| IPR policy, technology diffusion | Modifying IPR policy to provide for marching rights for social good when supported by public funds and for co-sharing IPRs generated under PPP; | Exploring newer mechanisms for fostering Technology Business Incubators (TBIs) and science-led entrepreneurship; | Providing incentives for commercialization of innovations with focus on clean technology | | |

Source: author's own elaboration

This table explains some of the major components of the STIP 2013, and factors for assessing water policy with respect to S&T objectives outlined in STI Policy. However, it is to mention here that additional empirical and theoretical concepts and research are needed to add rigour to understand barriers and challenges to understanding the innovation process in the realm of water policy reform in India. For instance, the research on S&T led innovation could be examined to determine what are the areas that could benefit from innovation in technology and which sectors are most frequently involved. This may answer questions such as whether hydrologists play a different role than industry experts, and if that is similar across sectors – essential information for any practitioner seeking to develop network relationships.

Table 4.5: Key components of STIP 2013 and their building blocks

Components of STI Policy 2013	BUILDING BLOCKS				
	Enabling framework	New structural mechanisms/ models	Regulatory mechanisms	Technology diffusion mechanisms	Impact assessment/mapping STI effectiveness
Investment in R&D/science-	More responsibilit	Modify decision	Reform phase and	Policy to be in	Percentage increase in R&D

Contd...

led technology	y than authority, strong oversight from DST or the govt.	making structures, develop mechanisms for leveraging academia-research-industry partnerships	the political context, co-investments of the public and private sector into R&D, parity investment by 2020	place that permits or promotes the diffusion of science-led technologies, apply more creative use of patents to gain added information and new ideas	investment, increase in academia-research-industry partnerships
S&T led innovation	world class infrastructure, finances, start-ups, establishment of "safe" spaces for innovation experiments	administrative capacity to address complex environmental issues	Support STI driven enterprises with business models, support combination of enterprises, universities and research institutes	sharing of IPRs between inventors and investors, bottom – up need based funding to areas affecting inclusive growth	Increase in number of start-ups, incubation centres, IPR between inventors and investors, increase in accessibility of globally competitive technologies
Integration of innovation for inclusive growth	Balancing collaboration and competition, social inclusion in technology choice/selection, S&T initiatives in conjunction with the less	ability of agents to integrate multilayered institutions, and diverse forms of power	Deployment of technology-led services for transparent governance	Linking to State Councils and NGOs, trigger mind set changes for value creation from knowledg	Integration of innovation for value chain creation, increase in access, availability and affordability of key S&T products and services

Contd...

	fortunate who will directly benefit by their empowerment			e	
Skill building /enhancing skills amongst youth	Enhancing skill for STI among the young through community colleges and polytechnique centres	Invest into young innovators through education, training and mentoring	Methods to make science attractive to the young, improving laboratory facilities for practical demonstrations		
Promotion of scientific temper	Leverage traditional knowledge through science	Stimulate research in universities and colleges, position performance linked reward system	More efficient information flows throughout civil society and the state apparatus		Effective science communication and deployment of National Knowledge Network
Short-term vision of global scientific power	Increase home-grown innovation and re-innovation	Assimilation and absorption of imported technology, collaborate internationally in mega-science	Develop systems for local S&T intelligence to maximize collection, analysis, dissemination and feedback at the global level		Establish a review mechanism to measure global performance. especially S&T activities within Ministries, Universities, IITs and main statutory bodies
National data centre and	creation or change of	creation of a new	regulations for data	develop network	New linkages between

Contd...

| data share system/mechanisms | an organizational structure, such as the development of a river basin organization or creation of a National Water Informatics Centre | multi-institution agreement | sharing and knowledge network, promoting collaboration and resource sharing between research institutes and universities | relationships | enterprises and knowledge centres |

Source: author's own elaboration

Some of the major water management challenges include stressed water basins where water allocation is equal to or more than the available water (surface water); extraction exceeding recharge (groundwater); rapid urbanization and industrialization which give rise to new priorities and demand in water use; and globalization which influences water policy and politics (Mollinga and Tucker, 2010). The summary of these challenges along with their supply and demand-side management is provided in the table below.

Table 4.6: Water management challenges, supply management and STI policy integration

Water management challenges	Supply Management	STI Policy integration
Water Planning i. Non Scientific ii. Fragmented iii. Delayed implementation because of disputes		S&T led innovation, Permitting multi stakeholders' participation in the Indian R&D system, River-basin as the planning unit, craft policies, strategies and plans to use S&T innovatively to tackle fragmented water planning, deepen and expand Information and Communication Technology (ICT) competences in water planning

Contd...

Rapid growth in demand of water due to i. Increasing Population ii. Increasing Urbanization, and iii. Changing lifestyle	**Creation of large storages and linkages** It is planned to create additional live storage capacity of 170 BCM by 2050. Completion of the storage projects under construction by 2025 would provide live storage of 63 BCM.	The R&D investment of India is less than 2.5% of the global investments (~$1.2 trillion as of 2009), which is under 1% of its GDP. For the water sector in India, there is a need of greater R&D investment with S&T led innovations to provide easy and accessible techniques for water resource development and use and also to replace the conventional and outdated technologies.
Variability of available water in both spatial and temporal due to i. Abnormal behavior of climate	Improving R&D to reduce ecological pressures, as well as, address the exigencies of natural and man-made hazards, including droughts and floods	S&T led innovation can help in modeling and simulating flash flooding scenarios in flood-prone areas, and aid in designing appropriate water management infrastructures, reservoir systems and more to develop new innovations for predicting, managing and sometimes even preventing events that might otherwise have a negative impact on a region's water supply
Groundwater withdrawal ii. Used without reference to dynamic recharge iii. Leading to unsustainable conditions and wastage of energy	**Aquifer Recharge:** This would require construction of percolation tanks, check dams, contour bunds etc. to saturate the catchment area and increase abstraction efficiency to 90%, and recharge efficiency to 75%. **Rain Water Harvesting:** This involves harvesting rain water in the watersheds and using it for micro-irrigation in rain-fed cultivated areas. This will increase the yield of various crops by 25-40%.	Managed aquifer recharge through leading edge technology, technology choice with Central Ground Water Board, developing and maintaining scientific skills at local levels
Existing Infrastructure i. Not	**Large scale rehabilitation of irrigation works:**	Encourage closer working relationship between the

Contd...

ii.	maintained leading to under utilization of resource Natural water bodies encroached upon	Such an intervention would require renovation, de-silting and setting up of management infrastructure for irrigation works, creating an additional potential of 5 mha. **Last Mile Irrigation Infrastructure:** This will set up the command area management structure and rehabilitate the system to bridge the gap of 9 mha (approx) between the irrigation potential created and that utilized. **Small Scale Irrigation Infrastructure:** Minor irrigation infrastructure projects, such as dams built closer to the community for using water during dry spells, will have a potential of irrigating 1.5 mha.	public andprivate sectors and between academia and enterprises, Attract appropriate foreign direct investments in smart irrigation technology and management
Growing Pollution of water sources due to i. Non-treatment of sewage ii. Industrial effluents iii. Non-point source pollution affecting (a) Availability of safe water (b) Aquatic systems and their biodiversity		Increase investment in urban water-treatment facilities, improve the technology, encourage zoning of industries and CETP, strengthen knowledge surveillance capabilities and partnership creating systems between innovators and investors
Wastage and inefficient use of water because of - Low public consciousness about		innovation andtechnological upgrading in reuse of wastewater, develop skills, technical knowledge and

Contd...

−Scarcity of resource −Economic value		organizational techniques, to make newtechnologies for water efficiency work properly
Lack of adequate trained personnel in following areas- −For scientific planning −In modern technologies −In Analytical capabilities −Use of information technologies −Consensus building amongst stakeholders		associated training of personnel undertaken on a continuing basis, more skills and knowledge to be acquired to build linkages and partnerships withrelevant institutions, skilled and educated workforces in analytical capabilities

Source: author's own elaboration

4.5 DEMAND MANAGEMENT IN WATER SECTOR AND S&T LED INNOVATION

Laser Levelling: Use of laser levelling equipment for quicker and better levelling of the fields will contribute to water saving and increase water use efficiencies, besides reducing energy used in pumping water.

Zero or Minimum Tilling: This technology involves direct planting of the crops without any or minimum tillage of lands. It not only reduces water use by 20-30%, but also reduces cost of cultivation, increases yield by 10-20% and decreases greenhouse gas emission.

Sprinkler or Drip Irrigation: Use of sprinkler or drip irrigation saves 20-40% of water and increases yield by 10-40%.

System of Rice Intensification (SRI): This envisages transplanting seedlings of lesser age with more spacing and less water application only at saturation size.

Land Surface Modification, Bed and Furrow Irrigation and Drainage: Bed and furrow Irrigation permit growing of crops on beds with less water, reducing chances of plant submergence due to excessive rain.

Biotic and Abiotic Stress Management: The objective is to encourage better management of plant stress by optimum use of pesticides and innovative crop protection technologies.

Improved Germplasm: This would increase yield potential by using higher yielding seed varieties that are best adapted for specific conditions.

Increased Fertilizer Use: This would involve increasing fertilizer use to reduce mineral exhaustion and improve yields in irrigated lands. The yield of all crops will increase by 25-50%.

Irrigation Scheduling: The objective is to determine the exact amount of water for application to the field as well as the exact timing for application. The yield of all crops will increase by 5-20%, saving 10-15% of water.

Piped/lined Water Conveyance from Tubewells: This reduces the losses in the conveyance system. Use of piped/lined water conveyance from tube-wells saves 20-40% water and increases yield by 10-40%.

Subsurface Drainage: A subsurface drain is a perforated conduit of tile, pipe or tubing, installed below the ground surface to intercept, collect and/or convey drainage water. The yield would increase by 20-30%.

Development of Water Technology Hubs: These hubs will be useful for benchmarking the available technologies to provide a clear picture of the benefits to private entrepreneurs.

Engaging Local Users in Water Management: All stakeholders, including members of the public, need to be given full opportunities to share their views and influence the outcome of water projects impacting them..

Strengthening Technology Diffusion Network: The technology diffusion network needs to be strengthened. To start with, each Krishi Vigyan Kendras should have a water technologist.

Policy

Private Participation: Private participation in development and management of water resources, especially in large industrial clusters, needs to be encouraged.

4.6 NWP 2012 AND THE NEED FOR ECOSYSTEM APPROACH

The NWP 2012 mentions that water needs to be managed as a common pool community resource held, by the state, under public trust doctrine to achieve *food security, support livelihood,* and ensure *equitable and sustainable development* for all (NWP, 2012, section 1.3 (IV). It also mentions that water is essential for *sustenance of ecosystem,* and therefore, *minimum ecological needs* should be given due consideration (NWP, 2012, section 1.3 (V).

NWP 2012 has outlined that "safe water for drinking and sanitation should be considered as pre-emptive needs, followed by high priority allocation for other basic domestic needs (including needs of animals), achieving food security, supporting sustenance agriculture and minimum ecosystem needs. Available water, after meeting the above needs, should be allocated in a manner to promote its conservation and efficient use. It also mentions that "all the elements of the water cycle, i.e., evapo-transpiration, precipitation, runoff, river, lakes, soil moisture, and groundwater, sea, etc., are interdependent and the basic hydrological unit is the *river basin,* which should be considered as the basic hydrological unit for planning.

Further, Section 3.3 mentions that "ecological needs of the river should be determined, through scientific study, recognizing that the natural river flows are characterized by low or no flows, small floods (freshets), large floods, etc., and should accommodate developmental needs. A portion of river flows should be kept aside to meet ecological needs ensuring that the low and high flow releases are proportional to the natural flow regime, including base flow contribution in the low flow season through regulated ground water use".

Section 8.1 specifically emphasizes on conservation of rivers and river corridors, with managing the storage capacities of water bodies and water courses and their associated wetlands, floodplains, ecological buffer and other areas required for specific aesthetic, recreational and societal needs in an integrated manner to balance the flooding, environment and social issues.

Table 4.7: NWP 2012 and implications for ecosystem approach in river basin management

Sl. No.	Section of the National Water Policy 2012	Implications for ecosystem approach in river basin management	What is missing or what is required?
1.	Section 1.3 (ii)	principle of equity and social justice to inform use and allocation of water	equity and social justice principles are not considered while allocating water
2.	Section 1.3 (v)	water sustains ecosystem, hence minimum ecological needs to be given due consideration	Ecological needs have to be quantified for each segment of a river/sub-basin
3.	Section 3.3	ecological needs of the river to be determined through scientific study with respect to natural flow regimes as characterized by low or no flows, small and large floods etc. A portion of river flows to be kept aside to meet ecological needs	Regulatory measures lacking for maintaining portion of river flows to meet ecological needs
4.	Section 3.4	rivers and other water bodies to be considered for navigation opportunities as far as possible and all multipurpose projects	navigation possibilities to be explored keeping in mind environmental impacts if any
5.	Section 5.5	inter-basin transfers of water to be considered	ecological and social implications of IBT have

Contd...

		on the basis of merits of each case after evaluating the environmental, economic and social impacts of such transfers	been, and continue to be, inadequately addressed
6.	Section 8.1	conservation of rivers and river corridors, with managing the storage capacities of water bodies and water courses and their associated wetlands, floodplains, ecological buffer areas	wetlands, floodplains, ecological buffer areas have to be looked collectively, however, there are very few cases where they have been analyzed while balancing the floods and water storages
7.	Section 10.1	rehabilitation of natural drainage systems while developing strategies to avert floods and droughts	inadequate attention to ecosystem perturbation
8.	Section 12.3.	community participation in water resources projects or services	large projects ignore community participation
9.	Section 12.4 and Section 1.3 (VII)	river basin / sub-basin as a unit for planning, development and management of water resources	departments / organizations at Centre and State have not been restructured to manage water resources at a basin or sub-basin scale, a holistic and inter-disciplinary approach is missing
10.	Section 12.5 and 12.6	appropriate institutional arrangements for each river basin	river basin authorities/institutions have not been yet operationalised, existing Acts may have to be modified accordingly to enable establishment of basin authorities, comprising party states, with appropriate powers to plan, manage and regulate water resource in the basins

Source: author's own elaboration

The management of water in our country is by several agencies, many of which have duplicated roles with other agencies (See Table 2). A review of the projects on river-basin reveals that none of the agencies have taken ecosystem approach in managing/developing water resources. The rivers in our country are looked primarily by irrigation department. They do not integrate very well the ecological principles of rivers, floodplains, wetlands, river-corridors, paleo-channels etc. in water resources development and allocation. Even premier agencies like Central Water Commission and Central Ground Water Board make plans in a fragmented manner without giving due consideration to ecological services.

Table 4.8: Agencies/Departments under the Ministry of Water Resources, their primary functions and whether they have integrated ecosystem approaches in water resources planning and management

Sl. No.	Agencies/Departments	Primary responsibility	Whether they have integrated ecosystem approaches in water resources planning and management?
1.	Central Water Commission (CWC)	initiating, coordinating and furthering, in consultation with State Governments, various schemes for basin-wise development of water resources; control, conservation and utilization of water resources for flood control, irrigation, navigation, drinking water supply and hydro-power development; all matters relating to the Inter-State water disputes	Inadequately to some extent
2.	Central Ground Water Board (CGWB)	carry out scientific studies to monitor and assess groundwater, groundwater exploration aided by drilling, Regulation of groundwater	No

Contd...

		development by CGWA	
3.	Central Water and Power Research Station, Pune	R&D support to projects in the areas of water and energy resources development and water borne transport	No
4.	Central Soil and Materials Research Station, New Delhi	field and laboratory investigations, research on problems in geo-mechanics, concrete technology, construction materials for irrigation and power	No
5.	National Water Development Agency (NWDA)	prepare feasibility report of the various components of the scheme relating to Peninsular Rivers development and Himalayan Rivers development	No
6.	National Institute of Hydrology (NIH), Roorkee	undertake, aid, promote and co-ordinate systematic and scientific work in all aspects of hydrology	Inadequately to some extent
7.	WAPCOS Limited	consultancy services for development of water resources, irrigation and drainage, electric power, flood control and water supply projects	No

Source: author's own elaboration

4.7 CONCLUSION AND OUTCOME

There is indeed a water crisis today – even though water is one of the world's most abundant substances, it is also fast becoming one of the most stressed resources. Critical challenges in the water sector in India pertain to both quantum deficit and quality deficit. Quantum deficit is due to significant loss of groundwater, increase in area under irrigation, and increase in demand due to population growth. Quality deficit is mainly because of contamination. Two main areas where STI can contribute significantly in the water sector are: access to safe water and sanitation system, and technology for adaptation against water-related disasters such as floods and droughts. To deal with this challenge, both R&D and technology led innovation should focus on steps to reduce wastewater, prevent runoff pollution in rivers and lakes, and purify water to make it drinkable and reduce the energy used in the transportation of water. Similarly, S&T led innovation can help in modeling and simulating flash flooding scenarios in flood-prone areas, and aid in designing appropriate water management infrastructures, reservoir systems and more to develop new innovations for predicting, managing and sometimes even preventing events that might otherwise have a negative impact on a region's water supply. There is a need for concerted efforts with „solution science' to find S&T led innovative solutions to address water challenges in India. Role of better STI integration within the water policy framework is critical for operationalising guiding principles of STIP 2013 in a holistic manner. Similarly there are many other sections of water in which S&T are playing crucial role to improve the situation. However, still we need some more innovative research and technology in this sector. The paper indicates that customized policy tools in water sector can be made using building blocks from STIP 2013.

Using better STI integration, within the current water policy reform, India can achieve various objectives of NWP 2012 by facilitating the flow of evidence-based data across central and state organizations responsible for collecting and analyzing information on various components of hydrological cycle,water use and ecosystems so as to provide a shared comprehensive view of water resources landscape, and demand management.

REFERENCES

1. Asian Development Bank (ADB). 2009. Water Resources Development in India: Critical Issues and Strategic Options. Asian Development Bank, New Delhi.
2. Benson, D., Gain, A.K. and Rouillard, J.J. 2015. Water governance in a comparative perspective: From IWRM to a ‚nexus' approach? Water Alternatives 8(1): 756-773.
3. Bharati, L., B.K. Anand, and V. Smakhtin, 2010. Analysis of the Inter-basin Water Transfer Scheme in India: A Case Study of the Godavari–Krishna Link, p.63-78.
4. Biswas, A. K. and C. Tortajada. 2010. Future water governance: problems and perspectives, International Journal of Water Resources Development, 26 (2) pp. 129–139.
5. Census of India 2011, House listing and Housing Census data highlights, Houses Household Amenities and Assets.
6. Census of India. 2011. Size, Growth Rate and Distribution of Population.
7. Central Ground Water Board report (CGWB) (2009). Government of India.
8. Central Pollution Control Board (CPCB). 2009[a]. Status of Water Supply, Waste Water Generation and Treatment in Class I and Class II Cities of India. Ministry of Environment and Forest (MoEF), Government of India.
9. Central Pollution Control Board (CPCB). 2009[b]. Status of Water quality in India-2009. New Delhi; MoEF Central Water Commission.
10. Cullet, P. 2009. Water law, poverty, and development – Water sector reforms in India. Oxford: Oxford University Press.
11. Cullet, P. and Koonan, S. 2011. Water law in India: An introduction to legal instruments. Delhi: Oxford University Press.
12. DST. 2003. Science and Technology Policy. Department of Science and Technology, Government of India, New Delhi
13. Environmental Law Research Society (ELRS) 2012[a]. A Concise Guide to Water Laws in Uttar Pradesh.
14. Environmental Law Research Society (ELRS) 2012[b]. Governing water in India, Review of Law and Policy Developments.
15. Frans W.A. Brom, Chaturvedi, S., Ladikas M. and Zhang W. 2015. Institutionalizing Ethical Debates in Science, Technology and Innovation Policy: A Comparison of Europe, India and China In: M. Ladikas et al. (eds.), Science and Technology Governance and Ethics, Pp. 9 – 23. Springer.
16. Groundwater Year book (2012-13). Central Ground Water Board, Govt. of India.
17. Hydro met Division, India Meteorological department (IMD).

18. Katyaini, S. and Barua, A. 2015. Water Policy at science-policy interface – challenges and opportunities for India, Water Policy, DOI: 10.2166/wp.2015.086.
19. Kulkarni, H., Shah, M. and Shankar, P.V. 2015. Shaping the contours of groundwater governance in India. Journal of Hydrology: Regional Studies, 4, pp.172-192.
20. Mollinga P. P. and S. P. Tucker. 2010. Changing Water Governance in India: Taking the Longer View, South Asian Water Studies, vol. 2, no. 1.
21. Moorea, Michele-Lee; Portena S. von der; Ryan P., Oliver B., and Julia B. 2014. Water policy reform and innovation: A systematic review, Environmental Science & Policy, 38, Pages 263–271.
22. Najeeb K Md. 2013. National Aquifer Mapping- Expectations and Challenges. Jour. Geol. Soc. India, 81, 294-295.
23. Narsimhan, T N. 2008. A note on India's water budget and evapotranspiration. Journal of Earth System Science. Vol 117. No 3. PP. 237- 240.
24. NRAA, 2013. Contingency and Compensatory Agriculture Plans for Droughts and Floods in India- 2012. Position paper No.6. National Rainfed Area Authority, NASC Complex, DPS Marg, New Delhi-110012, India: 87p.
25. Planning Commission. 2008. Eleventh Five Year Plan (2007-2012). Chapter 6. New Delhi: Oxford University Press.
26. Planning Commission. 2010. Mid-term Appraisal of the 11th Five Year Plan.
27. Poddar, R., Qureshi, M. E., and Shi, T. 2014. A Comparison of Water Policies for Sustainable Irrigation Management: The Case of India and Australia, Water Resources Management.
28. Saravanan, V. S. 2015. Crafting or designing? Science and politics for purposeful institutional change in Social-Ecological Systems, Environmental Science & Policy 53, Part B: 225–235.
29. Schoeman, J., Allan, C., & C. M. Finlayson. 2014. A new paradigm for water? A comparative review of integrated, adaptive and ecosystem-based water management in the Anthropocene, International Journal of Water Resources Development, 30:3, 377-390
30. UNICEF, FAO and SaciWATERs. 2013. Water in India: Situation and Prospects.
31. United Nations Development Programme. UNDP. www.undp.org.
32. United Nations Environment Programme (UNEP). 2012. The UN-Water Status Report on the Application on Integrated Approaches to Water Resources Management, UNEP, Nairobi, Kenya.

Chapter 5

Use of Geospatial Technology in Environmnetal Impact Assesssment

Ekwal Imam, Orus Ilyas & M. Mukhtyar Hussain

Abstract: The fast growth in population, urbanization and change in land use pattern in developing countries have resulted in damage of historical, biological, archaeological and aesthetical values. There are needs of a tool which can be capable of complex analysis of these damages and produce an alternative plan. Environmental Impact Assessment (EIA) is a tool which may fulfill these requirements. The conventional way of Environmental Impact Assessment study is a less accurate and more time consuming process. Geospatial technology like remote sensing, Geographical Information Systems, and Global Positioning Systems are the latest technologies which may produce much more accurate results and perform various geographic analyses even in complex situations. Consequently, this technology has become an essential component of the Environmental Impact Assessment (EIA) process. With the use of spatial techniques EIA has enhanced substantial viewing, movement, query, and even map-making capabilities. This paper focuses on discussing the application of geospatial tools in EIA and the effective analysis of the natural resources for developmental planning, policy formulation, and decision making.

Keywords: Environmental Impact Assessment, remote sensing, GIS, GPS and development

5.1 INTRODUCTION

The concept of environmental impact assessment (EIA) was first recognised in earth summit held at Rhio in 1992. Principle 17 of the Rhio declaration states that – "EIA as a national instrument shall be undertaken for the proposed activities that are likely to have significant adverse impact on the environment and are subject to a decision of a competent national authority" (UNCED, 1992). In 1996, Canter, defined EIA as the systematic identification and evaluation of the potential impacts (effects) of proposed projects, plans, programs or legislative actions relative to the physical, chemical, biological, cultural and socio-economic components of the environment. In another word, EIA is a management tool to be carried out before any project or major activity to ensure that it will not in any way harm the environment on a short term or long term basis (Erickson, 1994).

Environmental impact assessment applies to all projects that are expected to have a significant environmental impact and address all impacts that are expected to be significant. It includes broad public participation and significant administrative review procedure. It also provides information that helps in taking the decision and enforcing the project. There are two fundamental approaches for determining whether or not to conduct a comprehensive environment impact study for a proposed action. First one is the policy delineations, whereas, second one is to conduct a preliminary study. These may help in taking the decision that a comprehensive study wouldn't be required (Muthusamy and Ramalingam 2003). Therefore, any developmental project requires not only the analysis of the need of such a project involved but, it requires a consideration and assessment of the effect of a proposed development on the environment (Tayagi and Singh 2014).

The purpose of environment impact process is to identify the potential beneficial and adverse impacts of development projects on the environment and other aspects of social, cultural, and aesthetic values. EIA is carried out before any project or major activity to ensure that it will not harm the environment on a short term or long term basis (Murthy and Patra 2005).

EIA covers generation of baseline information on land cover, vegetation pattern, geomorphology, hydrogeology, drainage pattern, air, water, and noise quality, socio-economics, etc. to assess the possible impacts and feasibility of a project(proposed) activity. In addition, it is also useful in drawing up effective Environmental Management Plans (EMP), which includes catchment-treatment plans, compensatory afforestation activities, resettlement and rehabilitation activities, land reclamation and temporal monitoring of effectiveness of EMPs, etc.(Joao and Fonseca 1996). It is true that activities carried out for any industrial development have impacts on environment and it deteriorate natural resources, like land, water, and forests, as well as other forms of biodiversity, which provide basic needs for food, water, clothing, and shelter. Therefore, there is a need of a tool that capable of analysing these damages and provide alternative plan. The conventional way of EIA is more time consuming process and providing less accurate data. The advanced techniques of remote sensing and GIS in conjunction with Global Positioning Systems (Geospatial technology) provide fast way of EIA and offer environmentalists, developers, and planners the means they need for ensuring the safety of the population, sustainable management of available resources, and decision-making processes.

The geospatial technology is useful for conducting environmental monitoring and assessment with respect to land-use/land-cover analysis, environmental change-detection studies based on multi-temporal satellite data, predicting vegetation-cover loss following project implementation, mapping soil-erosion intensity over the project area, mapping environmentally/ecologically sensitive areas or hotspots, selecting potential sites for environmental restoration measures, dispersion of pollutants, terrain models used to estimate shadow regions, slope and aspect allocation, allocation of land for different resources and preparing comprehensive thematic maps for planners, decision-makers.

5.1.1 History of Environmental Impact Assessment

After Second World War, rapid industrialization and urbanization in western countries was causing loss of natural resources giving rise to concerns for pollution, quality of life and environmental stress. Consequently in early 60s, pressure groups were formed with the aim of getting a tool that can be used to safeguard the environment in any development. The United States of America (USA) became the first country to respond to these issues and established a National Environmental Policy Act in 1970 and formulated legislation on EIA. In 1972 a Conference on Environment was held in Stockholm, attended by United Nations. During the conventions EIA was formalized and aftermath all developed countries have environmental laws (Lee, 1995).

5.1.2 Environmental Impact Assessment in India

In India till 1080s developmental projects were implemented with very little or no environmental concerns. The environmental issues began when a national committee on environmental planning and coordination was set up under the 4th five year plan (1969- 1978). After 1980, clearance of large projects from environmental angle became an administrative requirement to the extent that the planning commission and the central investment board sought proof of such clearance, which were attended by the Department of Environment (DOE). In 1985, DOE was upgraded to the Ministry of Environment and Forest (MOEF). A major legislative measure for the purpose of environmental clearance was undertaken in 1994 when specific notification was issued under section 3 and rule 5 of the environment protection Act, 1986. On 27 January 1994, the Union Ministry of Environment and Forests, Government of India, formulated an EIA notification making Environmental Clearance mandatory for expansion or modernization of any activity or for setting up new projects listed in Schedule 1 of the notification. Environmental clearance for development projects can be obtained either at the state level or at the central level depending on certain criteria concerning the characteristics of the project. However, for most projects the consent must first be taken from the state pollution control board or pollution control committees (Verma, Kumar, and Tiwari 2016).

5.2 PROCESS OF ENVIRONMENTAL IMPACT ASSESSMENT

Environmental Impact Assessment includes several steps listed below:

a) **Screening:** EIA process starts with project screening. Screening is done to determine whether proposal requires EIA or not and if so, at what level of detail. The output of the screening process is generally in the form of a document called an Initial Environmental Examination or Evaluation (IEE).

b) **Scoping:** There is no need to carry out always exhaustive studies on all environmental impacts for all projects. Therefore, scoping is used to identify the key issues of concern at an early stage in the planning process (Ahmed and Sammy 1987).

c) **Baseline data collection:** The term "baseline" refers to the collection of background information related to biophysical, social and economic settings of proposed project area.

d) **Impact analysis and prediction:** The core of environmental assessment process is predicting the magnitude of impact and their significance if a project is going to be implemented. Therefore, Impact analysis and prediction development is done (Morris and Therivel 1995).

e) **Analysis of alternatives:** Analysis of alternative is done to establish the preferred or most environmentally sound and financially feasible option for achieving project objectives.

f) **Mitigation and impact management:** Mitigation is done to avoid and minimize predicted adverse impacts for environmental management plan.

g) **Environmental Management Plan (EMP):** An Environmental Management Plan (EMP) is a detailed plan and schedule of measures necessary to minimize, mitigate, etc. any potential environmental impacts identified by the EIA. Once significant impacts have been identified, it is necessary to prepare an Environmental Management Plan (World Bank, 1993).

h) **Environmental Monitoring:** Environmental monitoring is the measurement of environmental indicators over time within a particular geographic area. Monitoring should focus on the most significant impacts identified in the EIA.

i) **Environmental Impact Statement (EIS):** The final EIA report is referred to as an Environmental Impact Statement (EIS). Most national environmental laws have specified what the content of EIS should have.

j) **Decision making:** At each stage of EIA, decisions are made. These decisions influence the final decisions made about the EIA. The EIS is submitted to designate authority for scrutiny before the final decision.

k) **Effective EIA follow-up:** In practice, an EMP, which is submitted with the EIS report, should be used during implementation and operation of the project. The link between EIA process and project implementation stage is often weak especially in developing countries (Welford, 1996).

l) **Public hearing and involvement:** After the completion of EIA report the law requires that the public must be informed and consulted on the proposed development after the completion of EIA report

m) **Projects that may require Environmental Impact Assessment:** The following projects generally requires Environmental Impact Assessment are nuclear power and thermal power plants, highway, ports, airports and river valley projects, petroleum refineries, exploration for oil/ gas and chemical fertilizers, synthetic rubber, cement and asbestos products, mining projects and primary metallurgical industries dyes and integrated paint complex, storage batteries, distilleries, pulp, paper and newsprint and electroplating.

5.3 EIA USING GEOSPATIAL TECHNIQUES

Environmental assessment is a prediction that envisages the likely consequences of implementing developmental projects, designing the appropriate preventive for negative impacts, and enhancement of measures that have positive impact. It encompasses the generation of baseline information on land cover, vegetation pattern, geomorphology, hydrogeology, drainage pattern, air, water, and noise quality, socio-economics, etc. to assess the possible impacts and feasibility of a project activity. The adoption of advanced technologies like remote sensing and GIS provides accurate and synoptic spatial and temporal databases on these variables/factors. In addition, it is also useful in drawing up effective Environmental

Management Plans (EMP), which includes catchment-treatment plans, compensatory afforestation activities, resettlement and rehabilitation activities, land reclamation and temporal monitoring of effectiveness of EMPs, etc. Satellite remote-sensing data have been effectively used in site selection, loss of agriculture/forest lands due to project activities, route alignment for power grids/pipe lines, ecological monitoring of thermal power plants, assessment of mining impacts, submergence area studies, impacts on wetlands, etc (Fig. 1).

5.3.1 GIS Approaches in Environmental Impact Assessment:

It is said that the very first time GIS was used for EIA in the late 1970s for the preparation of an Environmental Impact Statement (EIS) for a dam on the river Thames (Muthusamy and Ramalingam 2003). In the area of EIA, GIS has yielded excellent results by combining the areas of each individual assessment case and overlapping them with satellite remote-sensing data. Thereby, one can detect where the changes are in the landscape and vegetation before and after development, and determine whether the results of the development match the original proposals

5.3.1.1 Following are the GIS approaches in EIA:
a) Initial environmental examination (IEE)

b) Monitoring and interpretation of baseline data

c) Assessing impacts (especially cumulative impacts)

d) Identifying and analysing project alternatives (geographic location/site selection,

e) Overall design and choice of technology

f) Mapping of data during monitoring and auditing

g) Helping in decision-making or policy formulation

h) Environmental impact auditing.

5.3.2 Remote Sensing:

Remote sensing is a scientific technology that can be used to measure and monitor important biophysical characteristics and human activities on earth (Jensen, 2000). Managing the earth's natural resources and planning future developments are important for acquiring accurate spatial information through GIS to create important layers of biophysical, land use/land cover and socio-economic conditions in a GIS database derived from analysis of remotely sensed data (Star, Estes, and Davis 1991). Remote sensing can be used to identify and isolate regional to sub regional objects/factors of significance in an EIA in a time- and cost-effective manner (Satapathy, Katpatal, and Wate 2008).

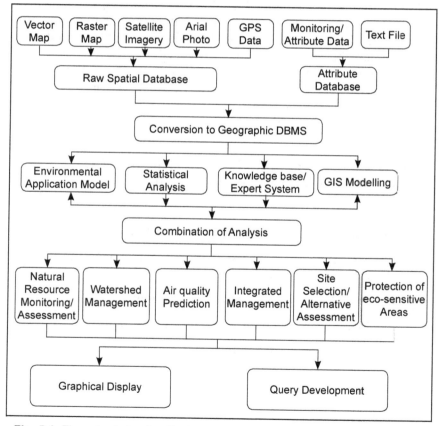

Fig. 5.1: Flow chart showing the integration of environmental datasets with geospatial technologies to assess environmental quality (Satapathy, *et.al.* 2008)

5.3.3 Environmental Impact Assessment Using Remote Sensing

5.3.3.1 Evaluation of urban development

Uncontrolled dynamic of the urban development leads to increase built-up surface areas. Negative action on the environment results from the lack of information on establishing a correct systematization and selection of areas for cities expansion. Using records of remote sensing with high spatial resolution has advantage in providing updated information on district and allows the estimation of systematic socio-economic parameters. Thus, remote sensing records provide tracking the extension of constructed surfaces. It is also used in conducting the general, urban and regional plans, in systematic planning of settlements and in any extension of localities (Weng, 2001).

5.3.3.2 Assessment in Mining

Mining operations involving the extraction of minerals from the area tend to have a notable impact on the environment, landscape and biological communities of the ecosystem (Bell, Bullock, Halbich, and Lindsey 2001). Unscientific mining pose serious problems, reducing the area covered by forests, soil erosion and pollution, air pollution, water and biodiversity loss. The conflict between mining activities and environmental protection has increased in recent years and it requires improving information on the dynamics of local and regional scale impacts (Latifovica, Fytasb, Chenc, and Paraszczakb 2005). Mapping and assessing impacts of mining are difficult problems because of the size and extent of areas affected by mines. Monitoring and controlling these activities through traditional methods is quite difficult due to high costs and lengthy time in obtaining accurate and updated maps. Satellite remote sensing together with geographic information systems are recognized as a powerful and efficient technology in assessing impacts on the environment caused by mining activities.

5.3.3.3 *Assessment of Changes in Land use/Cover:*

The changes in land use/cover lead to high social and economic benefits but they cost quite high on the natural environment. Studies on landuse/cover using remote sensing data have got a huge attention from countries around the world due to its importance in the overall analysis of change (Chilar, 2000) The actions produced by natural and human factors may lead to changes in terrestrial ecosystems (Houghton, 1994; Berlow, Bloomfield, and Dirzo 2000). in biodiversity and landscape ecology (Reid, Kruska, Muthui, Taye, Wotton, and Wilson 2000; Shi, Yuan, Zheng, WangJing-Ai, and Qiu 2007). It was found that for monitoring changes of land use/cover there is needed temporal resolution of at least 3-4 years. Even among the most developed countries possessing advanced equipment and detailed information on land use/cover, the lack of geospatial databases for landuse/cover can prevent the achievement of good planning practice. Therefore, satellite data are useful for both developed countries and developing ones where recent and reliable data on spatial information are lacking (Dong, Forster, and Ticehurst 1997).

Remote sensing data is used in assessing the environmental impact caused by changes and that may be recorded by very high resolution such as Ikonos and Ouickbird images, by medium spatial resolution such as Landsat (MSS, TM and ETM +) and SPOT images and by coarse resolution such as MODIS. Remote sensing helps in analysis of spatial assessments by mapping and classification of vegetation, evaluation of productivity of vegetation by vegetation indices, and studying the processes using the parameters obtained from satellite data. The analysis of changes in ecosystems and biomass requires information about crown such as leaf area index (LAI) and fraction of absorbed photo synthetically active radiation (Rahman, 2017).

Vegetation indices provide quantitative information on vegetation productivity based on spectral information that can be extracted from satellite imagery (Fig. 2). Thus, the AVHRR data with daily global coverage are used to calculate vegetation indices for monitoring land cover and vegetation phenology. Most widely used index of vegetation, NDVI index (Normalized Difference Vegetation Index), is linked to vegetation reflectance in red and near infrared bands and the factors affecting its values are vegetation cover density and the leaf chlorophyll of content. Other indices such as OSAVI (Optimized Soil-Adjusted Vegetation Index), MSAVI (Modified Soil-Adjusted Vegetation Index) and TSAVI (Transformed Soil-

Adjusted Vegetation Index) use different adjustment factors that SAVI index. Besides these indices based on ratios using orthogonal indices which depend on the soil line in spectral space most use are GI (Greenness Index) and GVI (Green Vegetation Index), (Vorovencii, 2011).

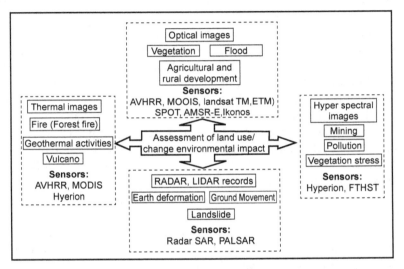

Fig. 5.2: Sensors used in assessment of land use/change environmental impact (Vorovencii, 2011)

4. USE OF GEOSPATIAL TECHNOLOGY IN EIA OF WILDLIFE

Disturbances to wildlife populations and their habitats is an inevitable outcome for earth system. No doubt EIA serves to enhance the conservation and management of wildlife through careful screening, scoping, prediction, mitigation and monitoring of project activities.

In recent years, models predicting species spatial distribution and habitat requirements have become prominent tools in wildlife management (Imam, Kushwaha, and Singh 2009; Syartinilia, 2008).The Habitat Suitability (HS) model is a perfect example of such a tool. Habitat suitability models predict the suitability of habitat for a target species based on its known affinities with environmental variables, consequently enabling the prediction of species spatial distribution (Hirzel, Gwenaelle, Veronique, Christophe, and Antoine 2006).

The use of HS models has been used for impact assessment for many decades, an example being its employment in 1983 by the U.S fish and wildlife service for Coho Salmon (McMahon, 1983). Considering that the increased availability of remote sensing data, GIS, and advances in statistical computation capabilities over the decades have permitted the development of more powerful techniques in habitat modelling. Therefore, HS model has been given considerable attention for EIAs where wildlife viability is at stake (Traill and Bigalke 2006; Syartinilia, 2008). In Sub-Saharan Africa, project alternatives presented for road development used HS maps to account for the habitat preferences of International Union for Conservation of Nature (IUCN) red listed species (Traill and Bigalke 2006). From such a model habitat suitability map can be produced. Lochran and Bigalke (2006) modelled the habitat suitability for large grazing African ungulates and suggested its utility for wildlife management. An output map of similar model can be seen below which predicts the spatial distribution of gaur in Chandoli tiger reserve (Imam and Kushwaha 2013).

In this study, habitat suitability index (H.S.I.) of gaur (*Bosgaurus*) was developed in Chandoli tiger reserve, India. For this, remote sensing and geographic information system were integrated with multiple logistic regression. Linear imaging self-scanning satellite-III (LISS III) imageries of Indian remote sensing satellite-P6 (IRS-P6) was acquired and digitally processed, whereas topographic maps were used for generating the collateral data in a GIS framework. Various layers of different variables such as forest density, Landuse land cover, measures of proximity to disturbances and water resources and a digital terrain model were created from satellite and topographic data. These layers along with global positioning system location of gaur presence/absence and MLR technique were integrated in a GIS environment for modelling the H.S.I. of gaur. The results indicate that approximately 31.14% of the forest of tiger reserve is highly suitable for gaur (Imam and Kushwaha 2013). Wilsey. Joshua, and David (2012) also used geospatial technology to evaluate performance of habitat suitability models for the endangered black-capped vireo

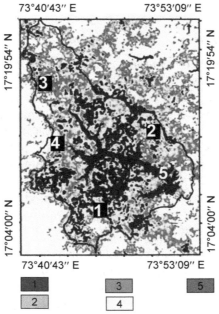

1: Highly suitable, 2: Suitable, 3: Moderately suitable, 4: Least suitable, 5: Water

Fig. 5.3: Habitat suitability map of gaur in Chandoli tiger reserve (Imam and Kushwaha, 2013)

During the prediction and mitigation stage of the EIA, HS models could be used to predict how species will respond to habitat fragmentation (Lochran and Bigalke 2006; Schadt, et. al. 2002). Habitat models can also be used identify and protect probable corridors between source populations, which is vital to ensure "species survival through maintaining genetic variability and providing a source of individuals to offset losses caused by poaching, predation and accidents (Gavashelishvili and Victor 2008). The relocation or repopulation of a disturbed wildlife species into a new suitable environment is another mitigation measure which can be achieved effectively with thorough data pertaining to species-habitat associations. Finally, HS maps can be useful for identifying areas where monitoring of wildlife viability should occur after the project has been implemented.

5.5 DISCUSSION

The data collected by remote sensing satellite pertaining to environment is very vast. Almost every year new systems go into operation, providing more advanced data. These technologies and data help in environmental monitoring and management decision-making. Over the past 20 years, sophisticated computer-based information systems have evolved that have great potential to help in developing management strategies for sustainable development and environmental protection. Geospatial tools are very useful in generating new information and in studying environmental change over time supported by raster GIS which are used for comparison, statistics, and presentation of findings.

EIAs are necessary tools for the planning, evaluation, and monitoring of sustainable development of the environment. Therefore, EIA needs to be carried out on all proposed projects, so that as infrastructure and economic development are enhanced, they will not be to the detriment of the environment. Remote sensing and GIS technology is a very useful method for carrying out standard EIA. It affords an easy and effective way of assessing impacts of projects on the environment, and also provides easy to understand maps, which could be readily studied by policy formulators.

Impressed with various international studies conducted to model habitat suitability index (Li, 1997; Hirzel, *et. al.* 2001; Wilsey, *et. al.* 2012), Indian researchers have also used remote sensing satellite data for habitat evaluation of various wildlife species like, Indian one-horned rhinoceros in Kazhiranga National Park, mountain goat in Rajaji National Park and tiger in Chandoli tiger reserve Maharashtra, (Kushwaha, Roy, Azeem, Boruah, and Lahan 2000; Kushwaha, Munkhtuya, and Roy 2001; Imam, *et. al.* 2009). Whereas, Imam (2005) evaluated habitat fragmentation indices by using FRAGSTATS computer software in remote sensing domain. Ministry of Environment, Forests and climate change (Govt. of

India) has developed national-level inventory and assessment of wetlands using LISS-III data.

Remote sensing plays important role in fire detection, its observation and its magnitude on ground. NASA's Fire Information for Resource Management System (FIRMS/GFIMS) was originally developed to get MODIS. It is reported that broadband, red-edge satellite information improves early detection of stress in a woodland ecosystem. Oil spill detection can be done by RADAR and thermal imagine. The radar based system has fully automated detection, giving oil spill position, tracking and measurement of drift (Imam, 2017).

5.6 CONCLUSION

To understand the consequences of human actions and natural phenomena on the environment, there is a need of real-time data for modelling various environmental impacts. Satellite remote sensing is an excellent tool where images with different spatial, spectral and radio metric resolution are used to fulfill the requirement. Use of satellite images in environmental impact assessment has the advantages of recording inaccessible and larger area in a single instance, repetitivity of the data, capturing of overall characteristics, and data can be retrieved at any time of day. In recent years, models predicting species spatial distribution and habitat requirements have become prominent tools in wildlife management. Habitat Suitability (HS) model is a perfect example of such a tool. It predicts the suitability of habitat for a target species based on its known affinities with environmental variables, consequently enabling the prediction of species spatial distribution.

REFERENCES

1. Ahmad, Y J. and Sammy, G.K. (1987). Guidelines to Enviromental Impact Assessment in Developing Countries, UNEP Regional Seas Reports and studies No. 85, UNEP.
2. Bell, F. G., Bullock, S. E. T., Halbich,T. F. J. & Lindsey, P. (2001). Environmental Impacts Associated with an Abandoned Mine in the Witbank Coalfield, South Africa. International Journal of Coal Geology. 45 (2). 195-216.

3. Berlow, E., Bloomfield, J. & Dirzo, R. (2000). Biodiversity: Global Biodiversity Scenarios for the Year 2100. Science. 287 (5459). 1770-1774.
4. Canter, L. (1996). Environmental Impact Assessment. USA: McGraw Hill Education.
5. Chilar, J. (2000). Land Cover Mapping of Large Areas from Satellites: Status and Research Priorities. International Journal of Remote Sensing. 21(6&7). 1093-1114.
6. Dong, Y., Forster, B. & Ticehurst, C. (1997). Radar Backscatter Analysis for Urban Environments. International Journal of Remote Sensing. 18 (6). 1351-1364.
7. Erickson, P.A. (1994). A Practical Guide to Environmental Impact Assessment. New York: Academic Press.
8. Gavashelishvili, A. & Victor, L. (2008). Modelling the habitat requirements of leopard Panthera pardus in west and central Asia. Journal of Applied Ecology. 45. 579-588.
9. Hirzel, H.A., Gwenaelle, Le Lay., Veronique, H.,Christophe, R. & Antoine, G. (2006). Evaluating the ability of habitat suitability models to predict species presences. Ecological Modelling. 199. 142-152.
10. Imam, E. (2005). Habitat suitability analysis of tiger in Chandoli National Park, Maharashtra using remote sensing and GIS [dissertation]. Dehradun: Indian Institute of Remote Sensing.
11. Imam, E. (2017). Remote sensing platforms and sensors (M-06). In: P-11. Remote sensing and GIS (GEL-11), e PG Pathshala, University Grant Commission (UGC), Ministry of Human Resource Development (MHRD), Govt. of India Available: http://www.epgp.inflibnet.ac.in/ January 10, 2017.
12. Imam, E., Kushwaha, S. P. S. & Singh, A. (2009). Evaluation of suitable tiger habitat in Chandoli National Park, India, using multiple logistic regression. Ecological Modelling. 220. 3621-3629.
13. Imam, E. & Kushwaha, S. P. S. (2013). Modelling of habitat suitability index for Gaur (Bos gaurus) using multiple logistic regression, remote sensing and GIS. Journal of Applied Animal Research. 41(2). 189 - 199.
14. Jensen, J. R. (2000). Remote sensing of the environment: An earth resources perspective. New Jersey, USA: Prentice-Hall.
15. Joao, E. M. & Fonseca, A. (1996). Current Use of Geographical Information System for Environmental Assessment: A discussion document. Environmental and Spatial Analysis. 36. Department of Geography, London School of Economics, London.
16. Kushwaha, S. P. S., Munkhtuya, S., Roy, P. S. (2001). Mountain goat habitat evaluation in Rajaji National Park using remote sensing and GIS. Journal of Indian Society of Remote Sensing. 28. 293-303.
17. Kushwaha, S. P. S., Roy, P. S., Azeem, A., Boruah, P. & Lahan, P. (2000). Land area change and rhino habitat suitability analysis in Kaziranga National Park, Assam. Tigerpaper. 27. 917.

18. Latifovica, R., Fytasb, K., Chenc, J.,Paraszczakb, J. (2005). Assessing Land Cover Change Resulting from Large Surface Mining Development. International Journal of Applied Earth Observationand Geoinformation. 7(1). 29-48.
19. Lee, N. (1995). Environmental Assessment in European Union. A tenth anniversary project appraisal. 7. 123-136.
20. Li W., Wang, Z., Ma, S. & Tang, H. (1997). A regression model for the spatial distribution of red-crown crane in Yancheng Biosphere Reserve, China. Ecological Modelling. 103. 115-121.
21. Lochran, W. T. & Bigalke, R. C. (2006). A presence-only habitat suitability model for large grazing African ungulates and its utility for wildlife management. African Journal of Ecology. 45. 347–354.
22. McMahon, T. E. (1983). Habitat suitability index models: Coho salmon. U.S. Department of Fish and Wildlife. Servo FWS/OBS-82/10.49.1-29.
23. Morris, P. & Therivel, R. (1995). Methods of environmental impact assessment. London: UCL press.
24. Murthy, A. & Patra, H. S. (2005). Environment impact assessment process in India and the draw backs. Environment Conservation Team, Vasundhara. 1-30.
25. Muthusamy, N. & Ramalingam, M. (2003, December 15-17). Environmental impact assessment for urban planning and development using GIS. In Proceedings of the Third International Conference on Environment and Health, Chennai, India.
26. Rahman, A. (2017). Environmental Applications of Remote Sensing (M-21). In: P-11. Remote sensing and GIS (GEL-11), e PG Pathshala, University Grant Commission (UGC), Ministry of Human Resource Development (MHRD), Govt. of India. Available: http://www.epgp.inflibnet.ac.in/ January 10, 2017.
27. Reid, R. S., Kruska, R.L., Muthui, N.,Taye, A., Wotton, S. & Wilson, C.J. (2000). Land-Use and Land-Cover Dynamicsin Response to Changes in Climatic,Biological and Socio-Political Forces:The Case of South-Western Ethiopia. Landscape Ecology. 15 (4). 339-355.
28. Satapathy, D. R., Katpatal, Y. B. & Wate, S. R. (2008). Application of geospatial technologies for environmental impact assessment: an Indian Scenario. International Journal of Remote Sensing. 29(2). 355-386.
29. Schadt, S. E., Wegand, T., Knauer, F., Kaczensky, P., Moser, U. B., Bufka, L., Cerveny, J., Koubek, P., Huber, T., Stanisa, C. & Trepl, L. (2002). Assessing the suitability of Central European landscapes for the reintroduction of Eurasian lynx. Journal of Applied Ecology. 39. 189-203.
30. Shi, P. J., Yuan, Y., Zheng, J., WangJing-Ai, Ge.Y. & Qiu, G. Y. (2007). The Effectof Land Use/Cover Change on SurfaceRunoff in Shenzhen Region, China. Catena. 69 (1). 31-35.

31. Star, J. L., Estes, J. E. & Davis, F. (1991). Improved integration of remote sensing and Geographic Information System: a background to NCGIA initiative. Photogrammtric Engineering and Remote Sensing. 57. 643–645.
32. Syartinilia ,T. S. (2008). GIS based modeling of Javan Hawk-Eagle distribution using logistic and autologistic regression models. Biological conservation. 141. 756-769.
33. Traill1, L. W. & Bigalke, R. C. (2006). A presence-only habitat suitability model for large grazing African ungulates and its utility for wildlife management. African journal of Ecology. 45. 347-354.
34. Tyagi , A. & Singh, V. (2014). Environmental Impact Assessment (EIA) Study of Non-Metal Mines: A Critical Review. International Journal of Engineering and Technical Research. 2 (5). 309.
35. UNCED. (1992, June 3-14). United Nations Conference on Environment and Development (Earth Summit) conference. Rio de Janeiro, Brazil.
36. Verma, A., Kumar, A. & Tiwari, V. K. (2016). Role of Remote Sensing and Geographical Information System in Environmental Impact Assessment of Developmental Projects for Environmental Management. International Journal of scientific research and management. 4(9). 4543-4553.
37. Vorovencii, L. (2011). Satellite Remote Sensing in Environmental Impact Assessment: An Overview. Bulletin of the Transilvania University of Braşov. 4 (53). 73-80.
38. Welford, R. (1996). Corporate Environmental Management, Earthscan, London .
39. Weng, Q. (2001). Modelling Urban Growth Effects on Surface Runoff with the Integration of Remote Sensing and GIS. Environmental Management. 28 (6). 737-748.
40. Wilsey. B. C., Joshua, J. L. & David, A. C. (2012). Performance of habitat suitability models for the endangered black-capped vireo built with remotely-sensed data. Remote Sensing of Environment. 119. 35-42.
41. World Bank. (1993). The World Bank and Environmental Assessment: An Overview. Environmental Assessment Sourcebook Update, No.1, Washington DC.

Chapter 6

Environmental Risk Assessment Regulation Of Genetically Modified Organisms

Dr. Faizanur Rahman

Abstract: *Genetically Modified (GM) Crops have their DNA altered i.e. a gene extracted from a living thing and placed in a different food by a scientist or an expert through genetic engineering to enhance desired traits such as increased resistance to herbicides or improved nutritional content. In contrary to conventional plant breeding methods, genetic engineering can create plants with the exact desired trait with great accuracy. However, the introduction of Genetically Modified (GM) crops into the environment and the food chain has become contentious issues in the India and else around the world. India adopted set of internationally accepted procedures of safety and risk assessment of GM crops that lead to the development of a regulatory system which protects human health and the environment and at the same time commands public confidence. The biosafety, risk assessment, agronomic evaluation, environmental impact and commercialization of GM crops involve carefully drawn guidelines, which are accepted under the Rules for the Manufacture, Use, Import, Export and Storage of Hazardous Micro-Organisms, Genetically Engineered Organisms or Cells in 1989 of the Environmental Protection Act 1986 of the Ministry of Environment and Forests. In this research work, the author will examine the regulatory structure in assessing ecological risk associated with the introduction of GM technology.*

Keywords: *Genetically Modified Organisms, Environmental Risk Assessment, Living Modified Organisms, Genetic Engineering Approval [Appraisal] Committee, Review Committee on Genetic Manipulation, Institutional Biosafety Committee*

6.1 INTRODUCTION

Biotechnology is an offshoot of science. Many definitions are available for it[1].Genetically Modified Plant and Organisms are being adopted globally for the past two decades. Now Genetically Modified Micro-Organisms are being transferred from the laboratory to the fields. Plants with new characteristics are being developed with the transgenic technology; for example, most of the genetically engineered crops are pest resistant, weed resistant, and so on. Thus, the farmers are reaping the benefits of this technology as there is now less dependence on pesticides. Another area of research involves developing protein-rich food having more shelf life. Agricultural biotechnology includes traditional breeding techniques that alter living organisms, or parts of organisms; improve plants or animals; or develop micro-organisms for specific agricultural uses.[2]However, modern biotechnology today includes the tools of genetic engineering.[3] Biotechnology increases productivity and reduces production cost. Genetically Engineered Plants are also being developed for phytoremediation in which the plants detoxify pollutants in the soil or absorb and accumulate polluting substances out of the soil so that the plants may be harvested and disposed of safely. In many countries, the debate on agricultural biotechnology revolves around the Living Modified Organisms (LMOs) only.[4]

The implementation of biosafety regime under the Cartagena Protocol on Biosafety (2000) is expensive and needs more experts and better technology. There is no scientific consensus on the free release of Genetically Modified Organisms (GMOs) and their impact on the environment. The governments and regulatory agencies around the world ensure a rigorous and thorough evaluation of GM crops for the environmental and ecological concerns prior to their release in the environment. Harmonized risk assessment procedures are adopted by different countries while considering the Environmental Risk Assessment (ERA) studies for the interactions between GM crops and its environment. The ERA obligates the developer of GM crops to generate safety data and information about the role of the introduced gene and the effects that it brings into the recipient plant soil biota, non-target organisms and environment. The risk assessment procedures require answers to specific questions about unintentional effects of GM crops on non-target organisms in the environment, whether the modified crop might persist in the environment longer than usual or invade new habitats and consequences of a gene being transferred unintentionally from the modified crop to other species.[5] Effective risk assessment and monitoring mechanisms are the basic prerequisites of any legal framework to deal with GMOs. In this context, it will be relevant to see how the Indian regulatory framework is structured and explore its weaknesses in view of the peculiar challenges posed by GM technology.

6.2 LEGAL SAFEGUARDS UNDER ENVIRONMENTAL LAW

Inspite of the reported benefits of GM crops, the apprehension mainly comes from an unknown risk that the new technology may bring. These concerns of biosafety have been adequately addressed in India by adopting set of internationally accepted procedures of safety and risk assessment of GM crops. The biosafety, risk assessment, agronomic evaluation, environmental impact and commercialization of GM crops involve carefully drawn guidelines, which are accepted under the Rules for the Manufacture, Use, Import, Export and Storage of Hazardous Micro-Organisms, Genetically Engineered Organisms or Cells in 1989 of the Environment Protection Act 1986 of the Ministry of Environment and Forests (MoEF).[6]

In India, the Guidelines, Protocols and Standard Operating Procedures (SOPs) were evolved over a period of time coinciding with the import, safety assessment, field testing and commercialization of Bt cotton from 1995 to 2002. The system of biosafety assessment and environmental impact assessment has been developed in conformity with the internationally accepted norms and protocols. The Rules 1989 established regulatory agencies such as the Institutional Biosafety Committee (IBSC), the Review Committee on Genetic Manipulation (RCGM) and the Genetic Engineering Approval [Appraisal] Committee (GEAC) to deal with the regulations of GM crops in the country. These agencies are assigned to the Department of Biotechnology and Ministry of Environment and Forests. A set of procedures have been laid out to regulate and comply with the requirements of R&D and commercialization of GM crops in the country. The Rules 1989 separately established full-fledged agencies such as the State Biotech Coordination Committee (SBCC) and District Level Committee (DLC) in collaboration with the respective State governments to undertake necessary steps for post commercial monitoring of GM crops. Over the years, it has been recognized that the effective functioning of the regulatory system requires transparency and accountability to instill public confidence in the system-a vital component for the success of GM crops in the country. With a decade of regulatory and field experiences with Bt cotton, it is well established that the existing regulatory framework is fairly robust and science based, however it failed to instill public confidence in executing its regulatory duties. Therefore, to keep pace with advances in modem biology and regulatory developments around the world, the Indian regulatory system on biosafety has to be revamped.

6.3 REGULATORY FRAMEWORK FOR GMOS

The basic legal framework governing GMOs (both GM crops and GM food products) in India is provided under the Environment (Protection) Act, 1986 (the 'EPA'). The central government formulated the Manufacture, Use, Import, Export and Storage of Hazardous Micro-organisms, Genetically Engineered Organisms or Cell Rules, 1989, which have been effective since 13 September 1993 (the 'Rules'). These Rules regulate all areas of research as well as large-scale application of GMOs and their products made in India, or imported into India. The Rules prescribe the requirements for risk assessment and regulatory approval for every proposed release of GMOs or GM products.

At the time when the Rules were formulated in 1989, there were no debates, specifically in the context of risks posed by GMOs. The source of the 1989 Rules, as stated in the preambular paragraph, is the power given to the central government under Sections 6, 8 and 25 of the EPA. Section 6 refers to rules to regulate environmental pollution, and specifically mentions the need for procedures and safeguards for the handling of hazardous substances. Section 8 mandates any person dealing with hazardous substances to comply with the prescribed safeguards. Section 25 grants the central government the power to make Rules, inter alia, for the purposes of prescribing procedures under Section 8. The Rules therefore seem to club GMOs with 'hazardous substances'.[7]

Research using GM technology in the early to mid-1990s led to greater specificity in the regulation of such technology under the Guidelines formulated under the Rules. The 1989 Rules were followed by the publication of the 1990 Recombinant DNA Safety Guidelines that were amended in 1994 (1990 Guidelines), and the Revised Guidelines for Research in Transgenic Plants, Guidelines for Toxicity and Allergenicity Evaluation of Transgenic Seeds, Plants and Plant Parts, 1998 (1998 Guidelines). Both Guidelines were formulated by the authorities in the DBT constituted under the 1989 Rules. The functions and responsibilities of the authorities are specified under the 1989 Rules as well as under the 1990 and 1998 Guidelines.

The Rules and Guidelines mandate, *inter alia*, the following:

i. Prohibition of unintentional discharge or release of GMOs.
ii. Prohibition of production, sale, import or use of substances and products including foodstuff, ingredients in foodstuff, and additives, which contain genetically engineered organisms or cells or micro-organisms, without the prior approval of the designated authorities.

6.3.1 Division of Jurisdiction under the Rules

Broadly, the Rules envisage division of jurisdiction, authority, and responsibility between the DBT and the MoEF, which has led to instances of both conflict and cooperation. The DBT was constituted under the Ministry of Science and Technology in 1986, for the general purposes of planning, promotion, and coordination of bio-technological programmes.

The 1989 Rules constitute regulatory committees under the DBT and the MoEF for the purpose of considering and giving approvals for GMOs for research and commercial use. While the DBT committees are responsible for considering GMO applications for research and small-scale field trials, the committee under the MoEF is responsible for large-scale trials and commercial use of GMOs. Applications for food safety before commercialization of food products containing GMOs are also considered by the authority under the MoEF. The distribution of powers and responsibilities among the various committees under the DBT and the MoEF is discussed in the following sections.

6.3.2 Authorities and Their Responsibilities

The 1989 Rules mandate the creation of six competent authorities, each having jurisdiction over a particular aspect of biotechnology. The RDAC (Recombinant DNA Advisory Committee) and the RCGM (Review Committee on Genetic Manipulation) are committees under the DBT. The RDAC is responsible for making recommendations on rules and procedures for ensuring biosafety in research and applications of GMOs. The RCGM is or for granting approvals for and monitoring safety aspects of research projects involving GMOs. It can also give approval for controlled field experiments.

The GEAC (Genetic Engineering Approval Committee) functions under the MoEF and is responsible for the approval of proposals relating to release of genetically engineered organisms and products into the environment in large-scale field trials. Given its broad mandate for granting approvals before the commercial release of all GMOs and their products, the GEAC is the authority responsible for food safety approvals for GM food products as well.

Every institution (both research institution and companies) undertaking biotechnology research is expected to constitute an IBSC (Institutional Biosafety Committee) as the nodal point for interaction within an institution for implementation of the guidelines. The IBSC is required to have a nominee from the DBT responsible for overseeing the activities of the institution.

The SBCCs (State Biotechnology Co-ordination Committees) are to be constituted at the state level, and have responsibility for the periodic review of safety and control measures in the various industries and institutions handling GMOs. The SBCCs function under the supervision of the GEAC at the MoEF. The DLC (District Level Committee) is to be constituted under the district collectors in every district of a state to monitor safety regulations in installations engaged in the use of GMOs.

The 1998 Guidelines introduced a seventh committee- the MEC (Monitoring and Evaluation Committee). It is authorized to conduct field visits at experimental sites, collect data on comparative agronomic advantages of transgenic plants, and advise the RCGM on risks and benefits, including recommending changes and remedial measures to the trial designs.

The composition of each authority varies, depending on the nature of its function. Broadly, they comprise officials from the relevant ministries of the Government of India, like the DBT, the MoEF, the Ministry of Health, the Ministry of Agriculture, and in some cases, independent experts nominated by the government.[8] Table 1 provides information on the composition of each committee.

Table 6.1: Composition & Committee

Authorities	Composition
RDAC Recombinant DNA Advisory Committee	The Committee comprises members nominated from the Department of Biotechnology
RCGM Review Committee on Genetic Manipulation	The Committee comprises representatives from the Department of Biotechnology, Indian Council of Medical Research, Indian Council of Agricultural Research Indian Council of Scientific and Industrial Research, Department of Science and Technology, and three experts in their individual capacity. The RCGM may appoint Subgroups to monitor specific projects
IBSC Institutional Biosafety Committee	The Committee comprises the head of the institutions, scientist in the institutional engaged in genetic modification technology, a medical

Contd...

	expert, and a nominee from the Department of Biotechnology.
GEAC Genetic Engineering Approval(Appraisal) Committee	The committee is to be chaired by the additional secretary from the department of Environment, Ministry of Environment and Forests and co-chaired by a representative from the Department of Biotechnology and Department of Atomic Energy, expert members including the Director General of the Indian Council for Agricultural Research, Director General of the Indian Council for Medical Research, Director General of Health Services, Plant Protection Adviser, Chairman of the Central Pollution Control Board and three outside experts in their individual capacity. The committee may co-opt members/experts as necessary

Table 2 provides in a nutshell the jurisdiction of the authorities with regard to various aspects of GMOs.

Table 6.2: Jurisdiction of the Authorities with regard to various aspects of GMOs

Name of Activity	Responsible Authority
Import/Exchange of GMOs or GM products (including Plants and food products) for research	Application to be submitted to RCGM for approval. In addition, phytosanitary certificates issued by the country of exports is required, which is to be routed through the NBPGR
Research for Development of r-DNA Products	Application to be submitted to RCGM for approval
Research for Development of transgenic plants	Application to be submitted to RCGM for approval
Environmental approval for large-scale use of food products, and	Application to be submitted to GEAC for approval. As a matter of

Contd...

clinical and veterinary processes containing GMOs	practice, the GEAC refers decision on health safety of GM foods to the ICMR
Environmental clearance for transgenic plants	Application to be submitted to GEAC for approval
Field visits to experimental sites and collection of data on comparative agronomic advantages of GM plants	Monitoring and Evaluation Committee of the RCGM

Both the RCGM and the GEAC follow the requirements of the 1990 and 1998 guidelines while considering a GM plant for approval. These guidelines provide the basic framework for risk assessment. For research activities, the guidelines have classified GMOs into three categories based on the level of the associated risk and requirement for the approval of the competent authority. The levels of risk and classification of the organisms within these categories have been defined in the guidelines. The guide-lines stipulate both physical and biological containment and GLP (Good Laboratory Practices). They also lay down criteria for GLSP (Good Large-Scale Practices) for using the recombinant organisms.

Depending on the nature and characteristics of the GM crop being evaluated for approval, the RCGM and the GEAC often design protocols on an ad hoc basis, specifying the parameters for testing, based on the peculiarities of each product being tested.

Detailed formats have been developed for eliciting information from an applicant seeking approval from the RCGM or the GEAC. A few experimental designs have been approved by the RCGM for conducting trials on GMOs in the open field. The formats for submitting applications/information to the IBSCs, RCGM and GEAC have been detailed out by the DBT, and are available on their website.

It has been observed that the regulatory and monitoring system instituted in India is amongst the most stringent and well developed in the world, and that the information sought is similar to the risk assessment models in other countries.[9] Evaluation of agronomic performance, ecological impacts, and health safety constitute the basic framework of risk assessment for both research and large-scale release. To summarize, under Rules, 1989, IBSC, RCGM-MEC and GEAC-ICAR/DCGI are involved in approval process of LMOs/GMOs and SBCC and DLC have monitoring functions. The procedures involved in the approval of GMOs in India are summarized below:

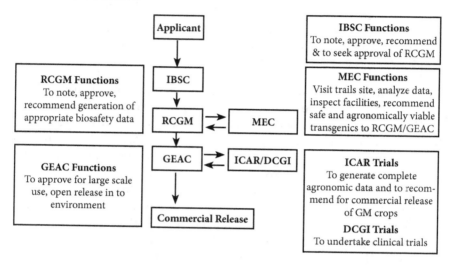

6.4 FUNCTIONING OF REGULATORY BODIES

The National Biotechnology Board, which was constituted in 1982 issued a set of biotechnology safety guidelines in 1983 to undertake biotech research in laboratory and contained use settings. In 1986, the National Biotechnology Board was promoted to a full fledge Department of Biotechnology (DBT) under the Ministry of Science and Technology.[9] In the early years, DBT monitored developments in the biotech field globally, developed safety guidelines and made efforts to promote large-scale use of indigenously relevant biotechnologies in the country. Realizing the importance of adequately assessing biosafety, biodiversity and environmental risks,[11] the research, product development and commercial release involving GMOs, hazardous microorganisms and trans-boundary movement of the living modified organisms (LMOs) by default were reallocated to MOEF. The Government of India (Allocation of Business) Rules 1961 assigned the responsibilities of 'biodiversity conservation' and 'environment protection' to MOEF in 1961.[12]

Thereafter, MOEF began regulating genetically modified organisms and products thereof under the existing Environmental Protection Act 1986which was enacted by the Parliament of India in 1986. Whereas the EPA 1986 does not describe GMOs and GM crops in the law per se, it lays down the legislative provisions to regulate 'hazardous substances' (Hazardous substance of the EPA 1986 section 3, 2 (iv) means any substance or preparation which, by reason of its chemical or physicochemical properties or handling, is liable to cause harm to human beings other living creatures, plants, microorganism, property or the environment) and to make administrative rules to regulate environmental pollution caused by hazardous substances. Henceforth, MoEF drafted and notified 'the rules for the Manufacture, Use, Import, Export and Storage of Hazardous Microorganisms, Genetically Engineered Organisms or Cells in 1989' (the 'Rules') under 'hazardous substances' section of the EPA 1986. As a result, GM crops, GMOs and the products of genetic engineering were de facto categorized as 'inherently harmful' in the same manner as hazardous substances that cause harm to human beings or other living creatures, property or the environment. Notably, the EPA Rules 1989 to regulate GMOs and GM crops were issued by an 'administrative order' through publication in the Gazette of India vide notification GSR 1037(E) dated 5 December 1989 and came into force vide notification S.O.677(E) dated 13 September 1993.[13]

The Rules 1989 cover a range of activities involving manufacture, use, import, export, storage and research of all genetically engineered organisms including microorganisms, plants and animals and products thereof. The Rules also apply to hazardous microorganisms that are pathogenic to human beings, animals or plants, regardless whether they are genetically modified. The Rules 1989 not only regulate research, development and large-scale commercialization of GM crops but also order compliance of the safeguard through regulatory approach, post-approval monitoring of violation and non-compliance.[14] The Rules define competent authorities and composition of such authorities for handling of various aspects of GMOs. There are six competent authorities that function into a three-tier system; the first tier includes the 'Policy Advisory Committee' such as the Recombinant DNA Advisory Committee (RDAC); the second tier consists of 'Regulating and

Approval Committees' such as the Institutional Biosafety Committee (IBSC), the Review Committee on Genetic Manipulation (RCGM) and the Genetic Engineering Approval Committee (GEAC), and finally, the third tier includes the 'Post Monitoring Committee' comprising of the State Biotechnology Coordination Committee (SBCC) and the District Level Committee (DLC). The functions of each of the committees have been articulated in the Rules 1989. In the spirit of inter-ministerial coordination, the implementation of the Rules 1989 was fast tracked with the subject matter experience and expertise of DBT. As a matter of fact, the Rules 1989 assigned the biosafety, risk assessment and risk management related aspects of GM crops to DBT. Notably, DBT was made an integral part of the Rules 1989, which was a unique feature allowing both ministries- MoEF and DBT of Ministry of Science and Technology-to regulate and safeguard from any foreseeable harm and weigh risks and benefits of GM crops and hazardous microorganisms under the Rules 1989. However, in case of post-monitoring of GM crops, the Rules assigned the responsibility to the respective State(s). It established a regulatory framework involving multiple government departments and delineated the administrative structure, authority, procedure and requirements for the regulation of GM crops at Union and States level. However, the Rules 1989 were unclear on the role and responsibility of the Ministry of Health and the Ministry of Agriculture-the important ministries that are empowered to regulate seed and human health and matters related to regulation of GM crops. Over the years, the biosafety regulatory system has evolved into a dynamic and comprehensive regulatory framework that involves different ministries; first, the ministries authorized under the Rules 1989 and second, the ministries that indirectly deal with GM crops. Figure1 describes the inter-ministerial coordinated regulatory framework on GM crops in India.

The regulation of GM crops from development, environmental release to commercial approval has been covered by three legislative Acts enacted by the Parliament of India and administered by different ministries. These included the Environment Protection Act 1986 implemented by MoEF, the Seed Act, 1966 & the Seeds (Control) Order by Ministry of Agriculture (MoA) and the Food Safety and Standard Act, 2006 (subsumed the Prevention of Food

Adulteration Act 1954) by the Ministry of Health and Family Welfare (MoH & FW).

The Rules 1989 were made central to the biosafety regulation of GM crops whereas others applied to food safety and quality of seeds for sale and matters connected thereto. The next layer of legislations (secondary legislation) dealt with import of material for R&D, access to biological resources and intellectual protection of plant varieties. Each Act has been implemented through a detailed guideline termed as Rules that described function, process, power and composition of different regulating agencies to implement the Act (Figure 6. 1).

As per the Rules 1989, the Recombinant DNA Advisory Committee (RDAC) set up by DBT brought out a first set of 'Recombinant DNA Safety Guidelines' in 1990 to regulate rDNA technology in medicine and agriculture. These guidelines were revised in 1994 as 'Revised Guidelines for Safety of Biotechnology'. Realizing the need for comprehensive guidelines for transgenic plants in the mid-nineties, DBT framed and released a comprehensive guide for GM crops in 1998 referred to as 'Revised Guidelines for Research in Transgenic Plants and Guidelines for Toxicity and Allergen city Evaluation of Transgenic Seeds, Plants and Plant Parts' to regulate GM crops and products. With regard to the application of GM crops, the guidelines 1990, 1994 and 1998 outline safety procedures, testing and use of genetically modified organisms and products. Considering the ecological consequences and the potential risks associated with the environmental release of GM crops, the guidelines prescribe the biosafety evaluation and risk assessment of the environmental aspects and agronomic performance on a case-by-case basis taking into consideration specific crop, trait and agro-ecological system.[15] These guidelines also call for regulatory measures to ensure safety of imported GM materials in the country.[16]

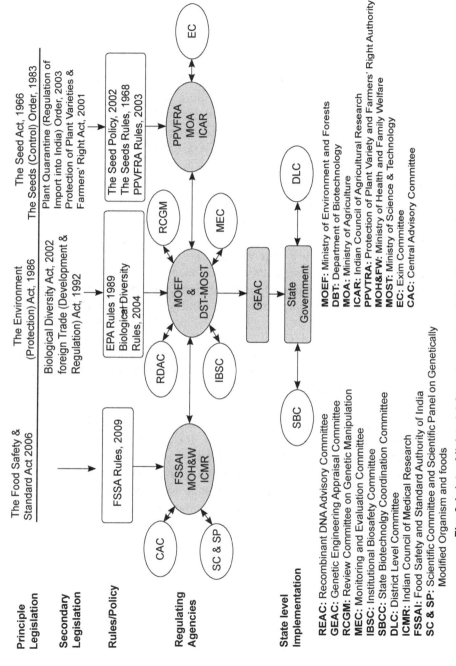

Fig. 6.1: Inter-Ministerial Coordinated Regulatory Frameworks on GM

6.4.1 Labeling of Genetically Modified Seeds

There is presently no law in India that specifically mandates labeling of GM seeds. However, labeling was mandated by the GEAC when it granted approval for Bt cotton. Labeling has been perceived as a logical requirement in order to ensure usage of the GM seeds for the purpose and the manner for which it is intended. Mahyco, the company that was given approval for the sale of Bt cotton seeds was mandated to label each packet of Bt cotton seeds to indicate the content and description of the seeds and provide detailed directions in the vernacular language for use of the seeds, including sowing pattern, pest management, and sustainability of agro-climatic conditions.

6.4.2 Monitoring Compliance

As discussed earlier, the 1998 Guidelines provide for a special MEC to be constituted by the RCGM to conduct field visits at experimental sites, collect data on comparative agronomic advantages of transgenic plants, and advise the RCGM on risks and benefits. Theoretically, the MEC can make these visits unannounced to the developer. However, in practice, visits are usually pre-planned and organized by the developer undertaking trials on the transgenic plants. The visits also last for a short period of time-ranging from a few hours to a few days. The visits pertain to only select locations of a field trial, usually identified by the private developer. Given the practical limits of field trials, it has been observed that these do not achieve a genuine monitoring and evaluation function, and rather serve to merely establish that the field sites actually exist.[17]

The SBCCs (State Biotechnology Coordination Committees) are meant to function under the GEAC with judicial powers to inspect, investigate, and take punitive action in case of violation of any of the provisions of the EPA. Specifically, they are responsible for reviewing, monitoring, and controlling the safety measures adopted while using GMOs in large scale in research, developmental, and industrial production within their jurisdiction. They are also responsible for overseeing field applications and experimental field trials. They are also supposed to provide information and data inputs to the RCGM on the safety and risks of approved projects. However, the SBCCs are not mandated to have any scientist engaged in transgenic research as their member, and questions have been raised about the competence of scientists to conduct monitoring and surveillance. Moreover, some states are yet to constitute SBCCs.

6.5 LIABILITY UNDER GMOS LEGISLATION

Liability for violation of the law and mechanisms for enforcement are the cornerstones of the regulatory process. The focus of the liability regime is criminal liability. The use of provisions relating to liability is therefore minimal. This section seeks to highlight the basic aspect of liability under the Environment Protection Act, the rules relating to GMOs, and jurisprudence of the Supreme Court of India in laying down certain principles of liability. It is being increasingly recognized worldwide that provisions for civil liability that mandate clear economic consequences for violation, could be a better mechanism for enforcement. The notified National Environment Policy, 2006, has highlighted this as an aspect that needs to be developed further in India.

6.5.1 Environment Protection Act, 1986

The penalty for non-compliance with any of the provisions of the 1989 Rules or Guidelines, or approval given under the same, is specified under the EPA. The EPA states that failure to comply with or contravention of any of the provisions of the EPA or Rules or administrative orders and directions issued under the EPA or its Rules (which includes the 1989 Rules) could result in imprisonment that could extend to five years, or a fine extending up to Rs. 100 000, or both.[18] In the event a company is guilty of contravention, every person directly in charge of the company can be deemed to be guilty of the offence and can be punished accordingly.[19] The presumption of guilt can, however, be rebutted by such person by showing evidence to the effect that the offence was committed without his/her knowledge, or that s/he exercised all due diligence to prevent committing such offence.

In the event the offence is committed by any department of the government, the head of the department shall be deemed to be guilty of the offence, unless s/he shows exercise of due diligence or that the offence occurred without his/her knowledge.[20] Any officer of the government conniving in or consenting to the offence or resulting in the offence due to neglect can also be deemed to commit the offence.[21]

Offences that could be prosecuted under the above-mentioned provisions would include (1) manufacture, import or sale of a GM plant or food product without the required permissions under the Rules and guidelines; or (2) contravention of any of the stipulations in any conditional approval accorded by the regulatory authorities.

The monetary limit of Rs. 1, 00,000 to the fine imposed under the EPA, may seem to be inadequate in view of the potential damage that may be caused by the failure to comply with the Rules, or order made under the Rules. Further, the EPA does not specify whether the violator of the Rules would be responsible for compensation for any damage caused.

Under the 1989 Rules, if an order under the Rules is not complied with, the DLC or the SBCC is empowered to 'take measures at the expense of the person who is responsible.[22] There is, however, no elaboration on what such measures could be. This provision is perhaps envisaged to cover possible compensatory and remedial measures to rectify the damage caused by the non-compliance with the Rules and orders made under the Rules, which would include all conditional approvals granted under the Rules. However, in the absence of any details on the scope and ambit of the SBCC's or DLC's power, any order by such authority regarding compensation is likely to be challenged before the judicial authorities. Another limitation is the lack of powers of enforcement vested with the authority.

It would, therefore, be trite to state that general liability rules under the EPA are inadequate to address potential liability issues in respect of GMOs. It would be necessary to develop separate liability rules for GMOs bearing in view the specific and novel challenges linked to the introduction of GMOs into the environment, including, but not limited to, illegal introduction of GMOs, escape of GM seeds, contamination of non-GM seeds, or failure of containment measures with regard to GMOs. Because of the uncertainties associated with

GMOs, the liability regime would need to constantly evolve and be amended to address future scenarios of risk and liability.

6.5.2 APPLICATION OF ABSOLUTE OR STRICT LIABILITY VIS-À-VIS TRANSGENIC CROPS

All the principles of liability discussed have been applied in the context of industrial environmental pollution and/or degradation. For example, the fact scenarios in the cases where the principles have been applied range from effluent discharge from a polluting industry, to industrial activity causing degradation of coastal zones, to air pollution from transport vehicles in cities. Visible 'environmental harm' has been a key feature in most of the cases. It would be interesting to see the applicability of these principles in a GMO scenario, wherein the harm is something associated with scientific uncertainty.

It could be argued that illegal cultivation or release of GMOs is an 'inherently dangerous' activity, or a 'non-natural use' of the land, with respect to which the principles of absolute or strict liability may be applied. The problem, however, is that the resultant 'harm' or 'damage' may not be immediately visible or quantifiable. The principles of absolute or strict liability (if applicable in a GMO scenario), therefore, need to be carefully articulated, to allow for scientific uncertainty and future damage or injury. It is, however, debatable whether such principles can be applied in the event of inadvertent escape or unanticipated damage caused by GMOs that have been released in compliance with statutory authorization.

Likewise, it is debatable whether the 'polluter pays principle' has relevance in a GMO context, unless illegal or unintentional release of GMOs and/or lack of containment measures are interpreted as 'genetic contamination', equivalent to environmental pollution.The 'precautionary principle', however, could have tremendous relevance in the GM context. The precautionary principle, so far, has been limited as a 'duty of the state' to take appropriate regulatory action, rather than as principle governing actions of private actors. It would, therefore, need to evolve further and be articulated more clearly in terms of its applicability in the GM context.

6.6 EVALUATION OF THE REGULATORY STRUCTURE

The moratorium decision on Bt brinjal by MoEF marked a considerable detour from an objective science-based rigorous institutional process of regulatory approval to a more subjective non-science driven decision-making process. This appeared primarily based on the high-pitched public consultations and selective invited interventions while ignoring the well-established regulatory and institutional norms of the country. Later, the notification issued on behalf of MOEF curtailed the role of GEAC from an 'approval' to an 'appraisal' committee and thereby systematically diminished the statutory authority of the former Genetic Engineering Approval Committee.[23] Apparently, the reasons were to effect the ministry's intention to subjugate approval power of GEAC and to some extent justify with retrospective effect, the ministry's controversial action to overrule the GEAC decision on Bt brinjal.[24] On the contrary, MOEF maintained that changing the name of GEAC does not undermine the authority and institutional framework of the regulatory system implemented by GEAC as per the Rules 1989.[25] This was the beginning of administrative and political interventions in the process of regulatory approvals of GM crops in the country.

These decisions have a far-reaching consequence on the overall functioning of the biotech regulatory system. It is important to note that GEAC is authorized to accord the approval of GM crops for import, field trials and release into the environment. The committee does not have a legitimate authority to grant commercial approval of GM crops. It is in this context that the environment ministry issued approval for environmental release of Bt cotton, whereas the seed licenses for commercial sale of each Bt hybrid were granted by the respective State(s) as per the provisions of the Seed Act 1966 and the Seeds (Control) Order 1983. The approval for commercial sale of Bt cotton seeds by respective State(s) was possible because the Indian constitution makes the environment the responsibility of the Union government whereas agriculture that of the respective State(s).[26]

In the case of Btbrinjal, GEAC did not accord the approval for environment release but concluded that 'Btbrinjal is safe for environmental release'. The committee forwarded the recommendation and report of the Expert Committee (EC-II) on the safety and efficacy of Btbrinjal event EE-I to MOEF. Subsequently, MOEF organized public consultations with a view to arrive at a decision to commercially release Bt brinjal,[27] which in fact overrode the legitimacy of the State governments to approve the commercial sale of seeds with or without biotech traits.

According to the Rules 1989, GEAC is the statuary committee with a sole mandate to regulate GM crops and accords approval for contained, confined and environmental release of GM crops in the country. GEAC is also entrusted with various provisions of the EPA Rules 1989 that require approval including:

i. Approval for the import of genetically modified microorganism for research purpose [Rule 7(1)].
ii. Approval for the use of hazardous microorganisms and recombinants in research and industrial production [Rule 4(3)(i)].
iii. Approval for all kind of experimental field trials of GM crops [Rule 4(3)(i)].
iv. Approval for measure concerning discharge of hazardous microorganisms [Rule 7(3)].
v. Approval for licenses for scaling up pilot project involving genetically engineered microorganism [Rule 7(4)].
vi. Approval for the deliberate release of genetically engineered organism [Rule 9(2)].
vii. Approval for certain substances containing genetically engineered organisms [Rule 10].
viii. Approval for food stuffs containing GMOs [Rule 11]
ix. Hold power to revoke any approval [Rule 13(2)].

Changing the name of GEAC from approval to appraisal committee contradicts the aforesaid provisions of the EPA Rules 1989. It also precludes the committee to exercise the authority to grant permission for approval of import, field trials and environmental release of GM crops in the country.

Moreover, in July 2011, MOEF required GEAC to adopt measures to issue permits to conduct field trials for research purpose only after the applicants submitted the 'no objection certificate (NOC)' from the respective State governments.[28] Unfortunately, there is no provision either in the Seed Act or the Seeds (Control) Order under which the respective State(s) can issue the no objection certificate for conducting field trials for research purpose.[29] Secondly, there are inadequate resources in the State(s) to assess and evaluate impending risk benefits of GM crops based on simple request from the applicant intending to conduct field trials of GM crops. Ironically, GEAC has been granting hundreds of field trial approvals of a dozen of crops from 1997 to 2011 without requiring a NOC from the States. The provisions of the EPA Rules 1989 have established a coherent mechanism to coordinate and monitor field trials by involving the respective State(s) through SBCC and DLC-State and District Level Committees. As a consequence of mandatory NOC, it triggers conflict between 'Union Vs States' resulting into an emergence of two sets of States, the ones that categorically refused to allow field trials and others setting up committee-after-committee to arrive at the decision of granting.[30] In the meantime, GEAC recognized that the frequent refusal of NOC by States was mainly due to lack of clarity and awareness on technical issues associated with biosafety measures.[31] For the last 2 years, the requirement for a NOC issued by the State(s) greatly complicated and delayed the approval process of field trials of GM crops.

In short, the present paper identifies three fundamental flaws in the current biosafety regulatory framework in the form of the EPA Rules 1989 that need to be rectified for the Indian regulatory system to function in a cost-effective and time-bound manner:

Firstly, GM crops are categorized as 'inherently harmful' under the 'hazardous substance' provision of the Environmental Protection Act 1986, which is scientifically incorrect and gives rise to misperceptions about the safety and potential risk of GM crops to health and environment.

Secondly, the EPA Rules 1989 to regulate GM crops were issued not by a 'legislative act' but by an 'administrative order' that remains untenable and liable to change with the desire of MOEF, which affects the predictability of the regulations and ignores the need to take into account the views and policies of other concerned ministries and the Parliament of India.

Finally, the Union environment ministry administers the regulation of GM crops in India whereas agriculture falls under the respective State(s). This often confronts approvals posing a 'Union versus State' conflict in decision-making on GM crops.

6.7 CONCLUSION

In conclusion, it is an imperative for the country to rely and guide the agricultural growth and food production with the help of scientific expertise and institutional capacity that have been built with painstaking efforts and enormous investment over a period of time. At no point in time should it endanger the institutional capacity of R&D, innovation and product deployment that are required to reverse the declining availability of food production. The new generation biotech traits, input traits in the short run and output traits in the long run have the potential to contribute to sustainable food, feed and fibre production. The development and deployment of novel traits would require the pooling of expertise and resources of both public and private sector institutions. Similarly, the regulatory processes have also to solicit the expertise of professionals and accredited laboratories from within and outside India to perform the regulatory oversight, stewardship and monitoring. Therefore, the current and future regulatory systems have to continue to harness the collective strength of interdisciplinary institutions to ensure the safety and efficacy of biotech crops. The current regulatory system of RCGM and GEAC has effectively utilized the existing expertise and capacity to build a system that regulates GM crops.[32] However, it remains untenable due to the fundamental flaws in the *Rules* 1989. The impact of the draft law on the *Biotechnology Regulatory Authority of India* (BRAI), which was recently introduced in the Parliament of India to create a new biotech regulator will largely depend on the extent to which it clarifies the Union's and States' jurisdiction and consolidates the decision-making responsibility in a non- politicized fashion, by adhering to the legislative parameters of purposefulness, clarity, transparency, predictability and efficiency. Adhering to the principles of federalism, the Government of India and the respective State(s) have to make major decisions to ensure the independence and proper functioning of the current and future regulatory system but not to be circumvented by rhetoric of precautions that may in the long-term jeopardize the national food security-a crucial part of the overall national security of 1.2 billion people.It is hereby recommended that no GM crop should be introduced in the country unless its effect on the environment and human health is scientifically assessed. This should be done by taking into consideration its long-term effects, and evaluation should be undertaken in a participatory, independent and transparent manner.

REFERENCES

1. The CBD (Convention on Biological Diversity), 1992 defines biotechnology as any technological application that uses biological systems, living organisms, or derivatives thereof, to make or modify products or processes for specific use. Life Science dictionary defines it as the industrial use of living organism or biological techniques developed through basic research. Biotechnology products include antibiotics, insulin, interferon, recombinant DNA and techniques such as waste recycling.
2. Biotechnology in the form of traditional fermentation has been used for decades to make bread, cheese or beer. It has also been the basis of traditional animal and plant breeding techniques, such as hybridization and selection of plants and animals with specific characteristics, to develop, for example, crops that gives higher yields of grain.
3. Gupta, A. K., & Chandak, V. (2005). Agricultural Biotechnology in India: Ethics, Business and Politics. *Int. J. Biotechnology*, 7, 212-227.
4. Under Indian law, the definition of Living Modified Organism will include only those organisms modified by r-DNA techniques through human interventions where the end product is a living modified organism.
5. Usha Rani, S., & Selvaraj, G. (2010). Performance of Bt. Cotton-Evaluation of Farmer Experience with Empirical Evidence. *Journal of Ext. Education*, 22(2), 4472-4479.
6. Raju, K. D. (Ed.). (2007). Genetically *Modified Organism-Emerging Law and Policy in India*. New Delhi: TERI Press.
7. Section 2(e), *Environment (Protection) Act*, 1986, Hazardous Substance means any substance or preparation which, by reason of its chemical or physico-chemical properties or handling, is liable to cause harm to human beings, other living creatures, plant, micro-organism, property or the environment.

8. Anuradha, R. V. (2007). Regulatory and Governance Issues Relating to Genetically Modified Crops in India. In K. D. Raju (ed.), *Genetically Modified Organisms: Emerging Law and Policy in India.* New Delhi: TERI Press.
9. Gupta, A. (2011). An Evolving Science-Society Contract in India: The Search for Legitimacy in Anticipatory Risk Governance. *Food Policy,* 36(6), 736-741.
10. Ghosh, K. (2008). Indian Efforts for Developing Biotechnology. *Asian Biotechnology and Development Review,* 11 (1), 35-56.
11. Ghosh, P. K. (1997). Transgenic Plants and Biosafety Concerns in India. *Current Science,* 72(3):172-179.
12. The Government of India (Allocation of Business) Rules, 1961.
13. Government of India. (1993). Ministry of Environment and Forest-Notification, S.O.677 (E). New Delhi: Ministry of Environment and Forest.
14. Ahuja, V. (2005). The Regulation of Genetically Modified Organisms/Food in India. In *Conference Proceedings 'Food Derived from Genetically Modified Crops: Issues for Consumers, Regulators and Scientists* (New Delhi: Indian Council of Medical Research).
15. Tripathi, K.K., &Behera, U.N. (2008). Agri-biotechnology in India: Biosafety Issues, Experiences and Expectations. In VibhaDhawan (Ed.), *Agriculture for Food Security and Rural Growth* (New Delhi: The Energy and Resource Institute).
16. Randhawa, G.J., &Chhabra, R. (2009). Import and Commercialization of Transgenic Crops: An Indian Perspective. *Asian Biotechnol. Dev. Rev.,* 11, 115-130.
17. Gupta, A., & Falkner, R. (2006). The Influence of the Cartagena Protocol on Biosafety: Comparing Mexico, China and South Africa. *Global Environmental Politics,* 6(4), 23-55.
18. Section 15, Environment *Protection Act,* 1986.
19. Section 16, *Environment Protection Act,* 1986.
20. *Id.,* Section 17(1).
21. *Id.,* Section 17(2).
22. Para 15, 1989 Rules
23. Gazette of India (2010). *Ministry of Environment and Forest-Notification,* GSR 613(E), the Gazette of India-Extraordinary, Govt. of India.
24. Editorial (2010). From Approval to Appraisal-Transparency and Professionalism needed in GM Policy. *Business Standard.*
25. Mukjerjee, S., & Menon, L. (2011, April 5). GEAC member quits over Conflict of Interest. *Business Standard.*
26. Choudhary, B. (2002). Indian Biotech sets a Constitutional Challenge. *Nature,* 419, 667

Chapter 7

Need for Environmental Information Policy: An Analysis of Environmental Impact Assessment for Proposed Integrated Steel Complex Site, Halakundi Village, Karnataka, India

Prof. (Dr.) K.M. Baharul Islam,
Archan Mitra & Dr. Asif Khan

7.1 INTRODUCTION

Rapid industrialization leads to the general development and improvement of people's standard of living, but without adequate environmental assessment the industrialization can become a disaster. To avoid such a precarious situation, we need the support of environmental impact assessment policies [i] laid down by the respective governments. Not only this, but the flow of environmental information[ii](Repository or database for environmental information, also known as the EIC) needs timely communication along with the pre and post assessment of both environmental and its social effects also needs to be carried out. Policy makers have framed policies based on reducing environmental pollution which is adversely affecting our planet, but the implementation and laying out policies are two different things which need to be synchronized. Synchronization requires an interface to work, when we discuss dissimilarity, we need to understand the interface that binds it. Today's Information Age[iii] has provided us raw material/tools with which we can work to create a plan of action to improve the information and communication interface amongst ourselves, which in turn will increase the entropy[iv] at all the level of communication. It will then be possible to create an information interface to bridge the dissimilarities and create a unison among the policy created by man for its natural environment. EIA works as an information interface between the officials of the government (the decision makers), society and the environment.

The paper is an attempt to analyze the environmental impact assessment report of Potpourri steel plant[v] and understand the information spectrum[vi] so as to propose a need for a comprehensive environmental information policy. India has a strong environmental policy, but access to information is still not at a significantlevel. Hence, environmental problems looms at large. The Environmental Impact Assessment (EIA) requires a spectrum of information from pre to post level of it. Thus, the researcher aims to identify the range of information, with the concepts of environmental information system[vii] and environmental informatics laid down to strengthen the proposed idea with the example of the EIA report of the steel plant in respect to information age today.

7.2 AIMS AND OBJECTIVES

The aim of the paper is to analyze the EIA report and find out the information spectrum that is the lifeline of the report. By finding out the spectrum, it was aimed to utilize it to propose the need for an information policy for EIA in India. It will further, help in the qualitative assessment of the regions that are taken for the EIA study. The new policy will comprise of ideas used in the EIS (Environment Information system) and EI (Environmental Informatics) to help work as an interface between the traditional approach and the new technocratic one.

7.2.1 Rationale of the Study

The paper is aimed to shed light on the lack of any institution to regulate information in our country. India does not have an Information Division, which can regulate the huge amount of data of information available digitally or in an analog format. This study tries to give a framework to the need for an information policy in one of the most diverse countries like India and specifically refers to the environmental sector. The researchers state the need for a working policy, including a working framework of how qualitative analysis is essential when we are taking a crucial decision based on only empirical judgment. Sometimes people's needs and wants are not adhered to along with their problems. The anthropogenic actions have always been non-reflective in such reports, based on which crucial decisions on policy matters are taken. This paper attempts to explore the environmental information while assessing the quality of the environment for the industrial purposes. It is a proven fact that we need to develop in tandem with our environment, if we are to create a sustainable way of living.

7.2.2 Limitations

This is a conceptual paper which includes analysis of one EIA Report, a more in-depth study needs to be placed with empirical data and evidence. As the report used is the older version, hence there is a need for a study into contemporary policies of EIA and Environmental Information dissemination in India and abroad. Due to temporal and spatial reasons, the researchers have been limited to the conceptual part of the study, thus focusing only on the needs for the development of an information policy.

7.3 METHODOLOGY

In this research paper, mixed method approachis utilized. The study has been done in different phases, and these are as follows:

Phase I: The first step of the research comprises of the case study in order to highlight the major points of importance in the EIA report.As a sample case report, the researcher has taken a single documentwhich has been presented to the Ministry of Environment and Forest (MoEF)byPopuri Steels Limited, the same was prepared by M/s. Global Enviro-Labs. The researchers,then analyzed the report to find out the information spectrum and also to identify the importance of environmental information which can be carried out further.For this phase, the researchers used a case study research design method. The convenience sampling method was used in selecting the case.

Phase II: The second phase of the research is done based on the literature review based on secondary sources like journals and books, whichprovided ideas of environmental informatics. It assisted researchers to propose a policy based on the information, apart from the policies which already exist. The researchers dealt with the needof information system in this phase after identifying the spectrum of information utilized in Phase I of the research. In this phase, they have used literature review and Meta-analysis process of the literature.

Phase III: The third phase of research has been done by reviewing the policies of MoEF laid down to formulate an EIA for a particular enterprise. Based on thereviews, the researchers came up with the findings to fulfill the research objective in the final phase.

PHASE I

IV. CASE STUDY ANALYSIS: POTPOURRI STEEL PLANT EIA REPORT

7.4.1 Overview of the Case

Karnataka is an exceptionally great state for mechanical development because of the accessibility of work, crude materials, and other framework. Steel being the basic product for every single mechanical action, the quantum of its utilization is considered as a record of modern prosperity. Since autonomy, there has been significant development in the steel area in India. The Ministry of Steel has set up different subgroups to help the Working Party on the Iron and Steel Industry for the Ninth Five-Year Plan and to 2006-07. As indicated by reports from the Ministry of Steel, add up to generation of completed steel was 20.75 million tons in 1995-1996 contrasted and 14.30 million tons in 1991-1992, demonstrating an expansion of 44, 8%. It was likewise anticipated that there would be 21.225 million tons of completed steel in 1996-97, 32.68 million

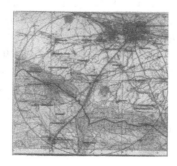

Figure 7.1

tons in 2001-02 and an expected 48.8 million tons in 2007-08. Accordingly, solid interest for steel in the modern, lodging and development areas was watched. To take care of the developing demand for steel, new steel plants must be set up. The steel enterprises, notwithstanding contributing essentially to financial development, likewise afterward, cause an ecological unevenness by making contamination. Hence, the exercises of the plant must be reasonable in its condition, with a specific end goal to diminish the natural effect of these exercises. To make a maintainable work design, we require a sound natural administration intend to be actualized by the plant's promoters, making ecological insurance an equivalent Production/benefit necessities (M/s Global Enviro Labs, n.d.).

The information that is utilized for the impact assessment is collected from various means available to the environmental engineers or lab personnel who are involved with the activities. Information about different elements affected by pollution is first monitored to make a status study of the area under concern. For this particular assessment, an area of the 10km radius is marked across the proposed site as shown in figure 1. Then the Environmental Impact Assessment study [1] is with itemized portrayal of different natural segments, for example, air, clamor, water; arrive, organic and financial inside a region of 10 km span around the proposed venture site ofMinesRoad, Halakundi Village, Bellary Taluk and District, Karnataka State according to the most recent rules of MoEF (Ministry of Environment and Forest)[viii],[ix] during the months of October 2008 to December 2008 (M/s Global Enviro Labs, n.d.).

7.4.2 Methodology Used In The Case
The diverse stages were associated with the natural effect evaluation think about and isolated into the accompanying stages:

   a. Identification of critical natural parameters and appraisal of the momentum condition of the effect region as for air, water, commotion, soil, organic and financial segments of the earth.

---

[1] This report has been formulated byM/s. Global Enviro Labs Tilaknagar "X" Roads, Bagh Amberpet, Hyderabad – 500 013. Ph.: 040 – 65582886; Telefax:040 – 27407969 E.Mail: Globalelabs@Rediffmail.Com for Popuri Steels Limited.

b. Prediction of an effect on air quality considering the anticipated emanations venture, the general situation.
c. Prediction of effect on water, arrive, natural and financial condition
d. Assessment of the aggregate outcomes in the wake of overlaying the situation accommodated in the standard situation to set up an ecological administration design. A programmed meteorological observing station to record meteorological parameters was introduced on the proposed venture site. Wind speed, wind heading, most extreme and least temperatures, relative stickiness and overcast cover were recorded every year finished a three-month time span from 1 October 2008 to 31 December 2008 Data. The breeze speed and wind bearing recorded amid the period. The investigation time frame was utilized for the count of the relative rate frequencies of the diverse breeze bearings. The gathered meteorological information was utilized to translate the current surrounding air quality and similar information was utilized to anticipate the effects of future situations because of the exercises of the proposed conspire. The measurement parameters for the data collection of the proposed site are presented in Table 1:

7.4.3 Analysis of the Methodology:
The data collection methodology of the EIA makes it explicit that the method was quantitative in nature. The qualitative information has not been utilized in the methodology of data collection for EIA. Environmental information is both quantitative and qualitative in nature, so just focusing on the quantitative parameters only gives us a wrong output. Thus, predictions of the future made on such assumptions are very harmful to the life of biotic and abiotic life near the area. It will so happen that, we would not be able to fully understand the extent to which the qualitative parameters can change over a quantitative period. This policy of only utilizing quantitative methodology calls for a change in reforms for data and information collection for the EIA reports. Because with more in-depth concepts of environmental informatics, we can bring change in patterns, identify problems more accurately and solve it with precision. Thus, the need for creating an information policy for EIA is the need of the hour.

Need for Environmental Information Policy...   157

Table 7.1: Parameters and Frequency of Data Collection for EIA Report

Attribute	Parameters	Frequency of Monitoring
Ambient Air Quality	SPM, RSPM	24 hourly samples twice a week for twelve weeks at eight locations
Ambient Air Quality	SO_2, NOx,	8 hourly samples average to 24 hrs for twelve weeks at eight locations.
Ambient Air Quality	CO, PAH	CO has been collected 4 hourly samples average to 24 hrs for twelve weeks at eight locations. PAH samples has been collected once at each station during the study
Meteorology	Surface: Wind Speed, direction, Temperature, relative humidity and rainfall	Surface: Continuous monitoring station for entire study period on hourly basis and also data collection from secondary sources.
Water Quality	Physical, Chemical and Bacteriological parameters	Once during the study period at eight locations
Biology	Existing flora and Fauna	Through field visit during the study period and substantiated through secondary sources
Noise Levels	Noise levels in dB(A)	Hourly observations for 24 hours per location
Soil Characteristics	Parameters related agricultural and afforestation potential	Once during the study period at eight locations
Land Use	Trend of land use change for different categories	By using Remote Sensing techniques and data from various Government agencies.
Socio-economic Aspects	Socio-economic characteristics, labour force characteristics, population statistics and existing amenities in the study area.	(Census Handbook, 2001)

Source: Census, 2001

7.4.4 Analysis of EIA

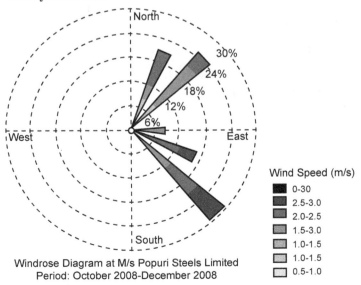

Windrose Diagram at M/s Popuri Steels Limited
Period: October 2008-December 2008

Figure 7.2

7.4.4.1 Ambient Air Quality

The encompassing air quality situation in the investigation zone was evaluated through a system of 8 surrounding air quality stations amid the examination time frame in a region of 10 km span around the undertaking region. The observing framework has been outlined such that agent tests are gotten from the heading of wind, wind and crosswind bearing of the undertaking site. These checking destinations have been built up considering accessible climatic models for wind heading and twist speed in this specific district. The accompanying focuses were likewise considered in the plan of the inspecting station organize:

- a. Topography/Terrain of the investigation range
- b. Areas populated in the examination range
- c. Residential and delicate regions in the examination range.
- d. Scope of encompassing businesses
- e. Representation of nearby baselines
- f. Representation of the cross-sectional circulation downwards (M/s Global Enviro Labs, n.d.)

To recognize air quality information and to speak to impedance from different mechanical and nearby exercises, screening methods were utilized to distinguish air quality stations in the investigation regions. The accompanying things were considered for the determination of air quality observing stations.

Predominant wind bearings.

- g. Topography of the investigative zone.
- h. Terrain and delicate regions.
- i. Populated ranges close to the undertaking region.
- j. Magnitude of the encompassing enterprises.

Table 7.2: Air profile Study

Locations	98th Percentile Values			
	SPM	RSPM	SO$_2$	NOx
Project Site	154	51	12.8	14.7
Halakundi	126	43	8.4	11.3
Mundirgi	115.5	31	6.2	8
Charakunte	113	40	6.30	8.5
Mineheri	112	36	5.2	5.9
Honnehalli	107.5	30	5.5	7.9
Bilematti	104.5	26.5	3.8	6.7
Honehalli tanda	106	28	5.8	7.2

A detailed study of air profile needs to be done;Table 2 shows the percentile of non-friendly air pollutants present in the atmosphere. Thus measures are to be taken not to increase the percentile of the gasses.

7.4.4.2 Noise Environment

Noise (Clamor) checking has been completed at eight areas to recognize the effect because of the current sources on the surroundings in the investigation range. Clamor levels were recorded at an interim of 30 minutes amid the day and evening times to process the day identical, night equal, and day-night parallel level. The acoustical condition shifts progressively in size and character all through generally groups. The clamor level variety can be fleeting, ghastly and spatial. The private commotion level is that level underneath which the encompassing clamor does not appear to drop down amid the given interim of time and is portrayed by unidentified sources. Encompassing commotion level is described by noteworthy varieties over a base or private clamor level. The greatest effect of clamor is felt in urban zones, which is generally because of the business exercises and vehicular development amid top hours of the day (M/s Global Enviro Labs, n.d.).

Table 7.3: Noise measurement Observation

Code	Location	dB(A)		
		Day Equivalent	Night Equivalent	Day-Night Equivalent
N1	Project site	51.6	46.3	49.1
N2	Halakundi	48.7	43.7	46.2
N3	Mundirgi	45.8	40.8	43.7
N4	Charakunte	46.7	40.1	43.5
N5	Mineheri	42.1	39.3	40.1
N6	Honnehalli	40.3	35.5	37.3
N-7	Bilematti	40.7	36.1	38.1
N-8	Honehalli tanda	41.6	35.2	37.2

Measured clamor levels showed as an element of time gives a valuable plan to portraying the acoustical atmosphere of a group. Commotion levels records at each station with a period interim of around 30 minutes are registered for comparative clamor levels. As appeared in Table 3. Equal clamor level is a solitary number descriptor for depicting time-changing commotion levels. The proportionate commotion level is characterized as numerically.

7.4.4.3. Water Environment

Eight groundwater tests from different areas around the venture site inside ten km sweep were gathered for appraisal of the momentum physico substance and bacteriological quality. Strategies received for examining and investigation were as per the IS techniques. Field parameters, for example, pH, Temperature were observed nearby. The parameters subsequently investigated were contrasted and IS 10500. The exercises encompassing the source amid inspecting were mulled over in the understanding of the water nature of that specific source.Numerous little streams are streaming southern way towards the Tungabhadra River. The examination zone does not have any human-made enormous water tanks with the exception of the characteristic lakes atHalakundi and Minehead. Because of the sparse precipitation in the course of the most recent couple of years, the tank and streams hold water just amid and soon after rainstorm season (M/s Global Enviro Labs, n.d.).

Table 7.4: Ground water Use

S. No.	Demand for	Ground water recharge (M cun/Annum)	Ground water consumption (M cum / annum)
1.	Industrial	–	0.211
2.	Domestic	–	3.56
3.	Agriculture	–	11.9
4.	Livestock	–	0.71
5.	Kvapotranpiration	–	1.752
	Total	175.212	18.133

Groundwater is the essential wellspring of water for the two water systems and also local purposes the investigation zone of 10 km. sweep. The groundwater potential in the range has not been abused to a most extreme degree, and there is a lot of extension for misuse. Dynamic groundwater assets are that measure of water, which is found in the common zone of change in an aquifer due to groundwater revive. The measure of groundwater revives and utilization.

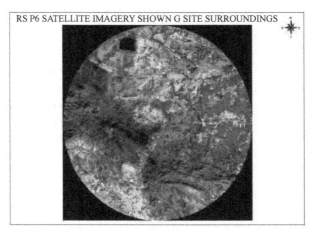

Figure 7.3: NRSA Satellite image of 10 km area for the proposed site

Land Environment: Ground truth ponders were led to distinguish the land use in and around 10 km sweep of the site. The agent soil tests were gathered from-eight examining areas inside a range of 10 km span around the proposed venture site for investigation of the physic-synthetic attributes to evaluate the editing design, microbial development, and so forth standard methodology were taken after for inspecting and examination. The examples gathered were additionally dissected to check the appropriateness for the development of local plant species in and around the task site.

Proposed development of the plant will occur in the region distinguished for the same. The required land has just been obtained. Land utilization of the region will change, and these progressions are permanent. The development of the proposed plant will upset the ground and soil strata, however the interruption is for genuine causes and on desolate land, and the effect will be permanent(M/s Global Enviro Labs, n.d.).

7.4.4.6 Biological Environment

Information on flora and fauna has been collected in the study area during the study periodwithin 10 km radius.Secondary data have been collected from various governmentdepartments such as forestry, agriculture; fisheries and animal husbandry establish the physiological, environmental conditions[x]. An ecological survey of the study area was conductedespecially in the reference torecording the existing natural resources. Secondary data was collected from forest department. The general vegetation along the road and agricultural lands are primarily due to the plantation of both exotic and native species of trees having commercial importance. Some of the commonly encountered species along the road-side are azadirachtinIndia, eucalyptus,acaciautriculiform's and prosthesisJuli flora. Domesticated animals of the area include cattle, buffaloes, goats, cats;dog's, etc. cows andducks are also common. No endangered faunal species are found in the area(M/s Global Enviro Labs, n.d.).

7.4.4.7 Socio- Economic Environment

The financial condition incorporates a portrayal of demography, accessible fundamental conveniences like lodging, human services administrations, transportation, instruction and social exercises. Data on the above-said factor had been gathered to characterize the financial profile of the examination range (10 km span), which is additionally a piece of ecological effect Assessment think about for the project.A definite financial study was directed covering all towns in the 10 km sweep from the middle. The data on financial perspectives has been assembled from different auxiliary sources, including different government and semi-government workplaces. (M/s Global Enviro Labs, n.d.).

7.4.4.8 Brief Summary & Analysis

As the Global Environs Lab report made the 2001 Census its base, therefore, according to it, there were 34 villages in the study area at that time. The population of the area was 72, 722 with a density of 231 persons per square kilometer.contrarily, the ratio of the female population is more than the Male population, 1000:953.The area was having inadequate educational facilities. Resultantly, the literacy rate of the area stands at a gloomy 38.75 percent, where male literacy was 25.69 percent and female rests upon miserable 13.06 percent. Theoccupational status of the study area rests upon agriculture, industrial andshortcommercial side. The people of the area mainly engaged in mine works with 47.16% of total population, where out of this percentage, 22.7 percent are agricultural labourers, as people engaged in agricultural activities in seasons only. A meager percentage of 5.10% also are marginal workers. The major crops in the area are groundnut,cotton, sunflower, etc. The medical facilities are inadequate, where only a few villages have the first health carecenters. For any major health care facility, the people of the area have to travel all the way to Bellary. Thecommon diseases are diarrhea, malaria, gastroenteritis and eye and skin diseases. However, there were bright side too, as many villages were adequately provided with potable water supply but on the other side, it was dearthof efficient sewage disposal system in the area.

The effect evaluation outline has both the positive and the negative perspectives set apart down, some of them being reversible and others not. The positive ramifications of the proposed action are normal amid the start-up of development exercises. Additionally, the nearby populace would have work openings in benefit exercises, contracts, and supply of building materials. Altogether, it would prompt a monetary upliftment of the area. Various specialized parts of the proposed venture have been concentrated to distinguish the noteworthy effects, which would emerge from the proposed movement. The recognized effects have has been measured through forecast of results to assess the post-venture scenario. Identified impacts due to proposed venture have been considered in detail to foresee the effects on different natural segments. The anticipated situation has been superimposed over the benchmark (pre-venture) status of natural quality to infer a definitive (post-project) scenario of ecological conditions. The Environmental Management Plan (EMP) for this proposed venture subtle elements the control measures, which will be reasonable for the proposed development to keep up natural quality inside as far as possible determined by State Pollution Control Board/CPCB/MoEF.

Table 7.5: Impact Characteristics of Environmental Attributes

Activity	Environmental Attributes	Cause	Impact characteristics			
			Nature	Duration	Reversibility	Significance
Site Clearing	Air quality (SPM and RSPM)	Dislodging of particles from the ground	Direct Negative	Short-term	Reversible	Low, if Personnel Protective Equipment (PPE) is used
	Noise levels	Noise generation from earth excavating equipment	Direct Negative	Short-term	Reversible	Low, if PPE is used by workers
	Land use	Residential land use	Direct Negative	Short-term	Irreversible	Low. As the proposed land use is non-residential
	Ecology	Removal of vegetation and loss of flora and fauna	Direct Negative	Short-term	Reversible	Low. No cutting of trees.
Transportation of construction materials	Air quality (SPM. SO_2, NO_x, CO)	Transport of construction material in trucks & Exhaust emission from vehicles	Direct Negative	Short-term	Reversible	Medium. Regular emission checks are performed
	Noise levels	Noise generation from vehicles	Direct Negative	Short-term	Reversible	Low, if regular vehicle maintenance is done.
	Risk	Risk of accidents during transit	Direct Negative	Long-term	Irreversible	Low, if safety measures are taken to prevent accidents.
Construction activities/ Laying of roads	Air quality (SPM, SO_2, NO_x, CO)	Operation of construction machinery, welding activities and others	Direct Negative	Short-term	Reversible	Low, if PPE is used by workers
	Noise levels	Noise generation from use of machinery	Direct Negative	Short-term	Reversible	Low, if PPE is used by workers
	Land use	Setting up of Project	Direct Negative	Long-term	Irreversible	The area is designated as Industrial area
	Ecology	Loss of vegetation	Direct Negative	Long-term	Reversible	Low. No cutting of trees and green belt development is envisaged

Need for Environmental Information Policy...

Table 7.6: Environmental Activities Matrix Table

Activities	Environmental Attribute								
	Air	Noise	Surface Water	Ground Water	Climate	Land & Soil	Ecology	Socio Economics	Aesthetics
Construction Phase									
Site Clearing	✓	✓				✓	✓		
Quarrying (indirect)	✓	✓				✓			
Ready-mix concrete preparation (indirect)	✓	✓				✓			
Transportation of raw materials	✓	✓							
Construction activities on land	✓	✓				✓	✓		
Laying of roads	✓	✓				✓	✓		
Operational phase									
Operation of DGs	✓	✓			✓	✓			
Solid waste disposal (indirect)	✓	✓	✓			✓			
Wastewater disposal			✓			✓			
Buildings									✓
Operation of Compressors		✓							
Operation of Vibrators. Conveyors		✓							
Vehicular Movement		✓							
Air Emissions from Stack and other Unit processes	✓								

Table 7.7: Impact Characteristics of Environmental Attributes

Activity	Environmental Attributes	Cause	Impact Characteristics			
			Nature	Duration	Reversibility	Significance
Afforestation/ Green belt development	Ecology	Planting of trees	Direct Negative	Long-term	Reversible	High positive impact
Emission from various unit processes and Vehicular traffic	Air quality (SPM, SO_2, NO_x, CO, HC)	Unit operations vehicle operation and fuel combustion	Direct Negative	Long-term	Reversible	Low as Ambient and Stack Monitoring vehicle maintenance, will be performed
	Noise levels	Noise generation from vehicles	Direct Negative	Short-term	Reversible	Low, with periodical maintenance of vehicles
Socio economic	Employment generation	Direct and indirect employment	Direct Negative	Long-term	Irreversible	High, new opportunities of steady income for many families
	Quality of live	In-flow of funds in the region/nation	Direct Negative	Long-term	Irreversible	High, the project will generate employment

Contd...

Solid waste disposal	Land and soil	Generation of solid wastes	Direct Negative	Short-term	Reversible	Low, proper collection and disposal
Waste water discharge	Water quality	Generation of waste water	Direct Negative	Short-term	Reversible	Low, as Septic Tank and soakpit will be provided
DG Set, Operation	Air quality	Exhaust emission	Direct Negative	Short-term	Reversible	Medium (DG set is only a stand by)
	Noise levels	Noise generation	Direct Negative	Short-term	Reversible	Low due to noise protection measures (enclosure, PPE etc.)

Findings from Phase I

From the EIA report, which is an environmental management plan for the proposed site, we can find that, quantitative information from the environment (Quality of air, soil, water and noise along with biodiversity and ecology). The data has been collected from both primary and secondary sources to help understand the status and condition of the area under study[xi].But the qualitative information has been overlooked during the assessment. Data and information gathering have been quantitative in nature and thus the results are also of the same nature. By discarding the qualitative aspect of the information, the conclusion that will be reached will not be a comprehensive one. EIA takes a lot of information and data as input variables to calculate the impact, by not following the qualitative/mixed methodology road, the result would not be the reality of the situation. There will be a research gap in the EIA and cause the GIGO (Garbage in Garbage out) phenomenon. This situation needs to be changed if we want sustenance, for that we must identify the spectrums through which,this information are traveling, and then only we will be able to calculate the information entropy (Shannon entropy).

Identifying the Information Spectrum through a Social Science perspective

The information spectrum is a subset of communication spectrum that deals with data. It helps is validating the information which flows on different channels/Spectrums to give a particular result(Han 2003). Here, nature\environment is the source of data and human beings utilize the data and information as spectrum to communicate with the natural sources. From this particular report, the researcher has identified the three major information spectrum from which the information as a spectrum flows from a particular state in a particular space and time.

Need for Environmental Information Policy...

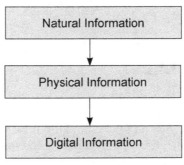

Figure 7.4

Natural Information Spectrum: Five elements of Nature (Naturally existing). Information about the elements of nature has been utilized as a major spectrum in the report.

Physical Information spectrum: Previous reports such as the census data and the topography maps, which had been prepared by the Geological Survey of India, constitute to be the physical information spectrum available to the people concerned with the report.

Digital Information Spectrum: Logistics and Computer mediated information\data processing. Data monitoring has been done by machine and computers to collect data, remote sensing and satellite imagery has been used an information spectrum.

The spectrum identifies that a constant flow of information is needed to predict the measurement and changes or fluxes in the environment. For example, atmosphere depends on thechangein precipitation, pressure, relative humidity, chemical composition, etc. If predictions are not made before implementation of any project that could bring change in the atmosphere, then it could be harmful to the environment. To help control such a scenario proper utilization of environmental information and social information needs to be done, the logical interfaces used in the air quality measurement are all quantitative in nature. Their qualitative effects on the lifestyle of the people and natural habitat need to be studied over time. To do so, utilizing information spectrum is of utmost necessity.

PHASE II

7.5 NEED FOR AN INFORMATION POLICY FOR EIA

5.1 Summary & Analysis
If we go through the summary of the analysis that has been done so far in the EIA, we can understand the information collected based on a small 10 km radius. The impact assessment matrix suggests that Noise and Air quality of the place under study are to be mostly affected by the industrialization. Other than that, land use and ground water level will be changed. Climate and ecology of the place havebeen demarcated as the least affected of all. If the data sets are considered, then it can be quantitatively proved the fact that these areas shall not be affected and even the impact on the affected areas shall be reversible. But if we consider qualitative environmental and social analysis of the place, we may find some constant changes that may occur. To do so, we have to know the proper use of EIS (Environment Information System) in EIA (Environmental Impact Assessment). The standard criteria for both the areas of EIA and EIS are information, whose importance has already been discussed previously in the paper with respect to EIA. Focusing on building a model for information, to protect the environment, support development through industrialization, and bring environmental and social sustenance. This model will be only possible if policy makers focus on making policy for environmental information which will support EIA, EIS, and all other information related industry/division shortly of our country in the information age.

5.2 Critical Analysis Environmental Information System
The age of information has many advantages. EIS has evolved significantly in the past 30 odd years or so. Technological innovations in computing have added to the change considerably (Haklay 1999). In a book from 1974 about EISs, the authors questions the need to computerize EIS and discusses other methods (Deininger and World Health Organization. Regional Office for Europe 1974). By 1998, the computer is seen as the only possible container for EIS (Gunter 1998). What the idea is rapidly changing, it is becoming more technocratic in nature, which is creating tension

in maintaining the balance of both traditional and modern applications to EIS. This balance is affecting the EIA to a great extent because the choice of technology is a critical aspect of any management and traditional EIS has shifted from balancing act to data sets. As it had previously been suggested that there is a need for new and improved information policy to deal with the data and information which are digitized and belong to the spectrum of digital information. GIS (Geographical information system) is closely related to the EIS, because when we follow the EIS we utilize the geographic information as well, which again belong to both the physical and digital spectrum as denoted by the author in the previous phase.As shown by Checkland and Holwell (1998) (cited in Haklay 1999) the result of this confusion may be the construction of small systems. EIA requires change and secondary policies. History shows a continuing trend to create specialized systems for use by experts. They are now faced with the difficult task of sharing the data, information and knowledge they store for the public. There is not a simple and easy technical solution to this problem, as this problem is mainly non-technical. One possible solution is to explore the development of the centralized public. In addition, bottom-up EIAs will help the public to use environmental information (p.19).

Since the early 1990s, a new area of research has been created for research on EIAs, coined the word *"Ecological Informatics"*. An abbreviated definition "Environmental computing is the domain of development, management and Research on environmental information systems"

"Environmental informatics is a field of applied computer science that develops and uses the techniques of information processing for environmental protection, research, and engineering[xii]. ... basic methodological issues and typical applications across a wide range of topics, including monitoring, databases and information systems, GIS, modeling software, environmental management systems, knowledge-based systems, and the visualization of complex environmental data." (Avouris& Page 1995).

The study of environmental informatics has helped in understanding the complexities that are created with excessive data inclusion. Hence, it can be said that present environmental informatics is the answer to the makethe EIA reports more effective and viable for sustenance, but before that it requires an Environmental Information Policy in India to support the concepts of the Environmental information system and Informatics.

PHASE III

7.5.3 Reviewing the Environmental Information System policies & EIA:

MOEF has 38 divisions depicted towards the protection of environment and forests among one of them is ENVIS (Environmental Information System). Infrastructure leasing and Financial Services (IL&FS)[2] had approached the MOEF with theidea of a repository of environmental information and was developed as EIC (Environmental Information Centre) with apartnership with MOEF. Though there are strong divisions in our Indian ministry, still there are no actions taken to develop policies based on information, Policy building for information can bring environmental sustenance. Humans today have the power of information and utilizing the power of information for protecting the environment is one of the most important tasks that man in the information age has the power to do. To help this progress, an environmental information policy needs to be laid down to support the EIA system and all the other systems that are entitledto it. The government of India is keeping no stones unturned to help support environmental information, which can be proved with the figure 5 and 6(Ministry of Environment, Forest and Climate Change, n.d.).

7.6 CONCLUSION

Though facts can be placed simply, but it is obvious that the information is very crucial to the matters related to the environment. Though it is deemed necessary and important by the people, still it is neglected by authorities when planning for the environment. In the cycle of decision making for a new policy, information should be

---

[2] Allahabad High Court makes DND Flyway toll free." Insert Name of Site in Italics. N.p., n.d. Web. 18 Nov. 2016 <http://newstrack.com/uttar-pradesh/allahabad-hc-dnd-flyway-toll-free>.

given priority. As of now, only data based on facts and quantitative information have been used, what the authors want to propose is the need for proper chain and system of information (Information Modeling) which will help in coordinating the messages and will ease the communication about the environment for policy makers to understand. The EIA has a policy of its own and is followed precisely by the agencies that are given the task of making reports. The problem is that these reports are static in nature after they have fulfilled their purpose there is no further use for it. Information from EIA reportsshould be used to reframe policies, not only based on mere facts or data, but also based on quantitative information as well. The need for environmental information policy is eminent, and the Indian government is taking a considerable amount of steps to prioritize EI through ENVIS and several other divisions. The Information Policy for the environment can only be visualized if we have productive information and communication systemin our country. It shall pave the way for sustenance in developmental ventures taken by the government of India. Development and environmental protection have been dissimilar in many respects, but by figuring out the information spectrum and utilizing them in devising the environment information policy for ENVIS and implement them while formulating EIA reports for developmental ventures can help solve the most complicated problem of development and environment to support sustenance.

REFERENCES

1. Deininger, R.A, WHO. Design of Environmental information system. Ann Arbor Science Publishers, 1974.
2. Gunter, O. "Environmental Information System." Springer 1998.
3. Haklay, Muki. From Environmental Information system To Environmental Informatics Evolution and Meaning. University College London. London: CASA, UCL, 1999.
4. Han, Te Sun. Information Spectrum Method in Information Theory. Berlin: Springer, 2003.
5. M/s Global Enviro Labs. Draft Environmental Impact Assessment Report For Proposed Integrated Steel Complex. Hyderabad: M/s. Global Enviro Labs, Hyderabad, 2009.

6. Ministry of Environment, Forest and Climate Change. Environmental Information System (ENVIS). 23 3 2016. 28 3 2016 <http://envfor.nic.in/division/environmental-information-system-envis>.
7. N.M Avouris, B.Page. Environmental Informatics: Methodology and Applications of Environmental Information Processing. Boston: Kluwer Academic, 1995.
8. P. Checkland, S.Holwell. Information, Systems and Information System - Making Sense of the field. Chichester: John Wiley & Sons Ltd, 1998.
9. Deininger, R.A. WHO. Design of Environmental Informatics Systems. Ann Arbor Science Publishers, 1974.

END NOTES

iEnvironmental Impact Assessment (EIA) is an important management tool for ensuring optimal use of natural resources for sustainable development. A beginning in this direction was made in our country with the impact assessment of river valley projects in 1978-79 and the scope has subsequently been enhanced to cover other developmental sectors such as industries, thermal power projects, mining schemes etc. To facilitate collection of environmental data and preparation of management plans, guidelines have been evolved and circulated to the concerned Central and State Government Departments. EIA has now been made mandatory under the Environmental (Protection Act, 1986) for 29 categories of developmental activities involving investments of Rs. 50 crores and above.

iiWith the greater need for accessing data for conducting an EIA or any kind of environmental assessment, availability of data in a ready and interactive format is the basic requirement. IL&FS first conceived the concept of evolving a knowledge database for this purpose. The concept of "Environmental Information Centre" (EIC) was strategized and developed in 2002 by IL&FS in partnership with the Ministry of Environment & Forests.

iii The Information Age (also known as the ComputerAge, Digital Age, or New Media Age) is a period in human history characterized by the shift from traditional industry that the Industrial Revolution brought through industrialization, to an economy based on information computerization.

iv In information theory, systems are modeled by a transmitter, channel, and receiver. The transmitter produces messages that are sent through the channel. The channel modifies the message in some way. The receiver attempts to infer which message was sent. In this context, entropy (more specifically, Shannon entropy) is the expected value (average) of the information contained in each message. 'Messages' can be modeled by any flow of information.

v "VALUES IN LITERATURE AND THE VALUE OF LITERATURE 12-14 ..." N.p., n.d. Web. 18 Nov. 2016 <http://blogs.helsinki.fi/values-in-literature/files/2012/05/Abstraktit-blogiin-v>.

vi Information spectrum, provide a very simple formalization and intuition into channel capacity in almostevery communication situation including unicast, multiple access, broadcast, and other situations.

vii Environmental Information System (ENVIS), by providing scientific, technical and semi-technical information on various environmental issues since its inception in 1982-83 (Sixth Plan), has served the interests of policy formulation and environment management at all levels of Government as well as decision-making aimed at environmental protection and its improvement for sustaining good quality of life of all living beings. The purpose has been to ensure integration of national efforts in web-enabled environmental information collection, collation, storage, retrieval and dissemination to all concerned, including policy planners, decision-makers, researchers, scientists and the public.

viii The Ministry of Environment, Forest and Climate Change (MoEFCC) is the nodal agency in the administrative structure of the Central Government for the planning, promotion, co-ordination and overseeing the implementation of India's environmental and forestry policies and programmes.

ix "Geospatial analysis to assess the potential site for coal...",N.p., n.d. Web. 18 Nov. 2016 <http://www.pelagiaresearchlibrary.com/advances-in-applied-science/vol3-iss3/AASR>.

x "Economics of Change in Cropping Pattern in Relation to ...",N.p., n.d. Web. 18 Nov. 2016 <http://ageconsearch.umn.edu/bitstream/204516/2/05-Subrata%20Kumar%20Ray.pdf>.

xi "Solenoid Valves Market by Technology - 2020 | IndustryARC.",N.p., n.d. Web. 18 Nov. 2016 <http://industryarc.com/Report/1273/global-solenoid-valves-market-analysis-report>.

xii "Environmental Informatics - Methodology and Applications ...",N.p., n.d. Web. 18 Nov. 2016 <http://www.springer.com/us/book/9780792334453>.

Chapter 8

Communication for Climate Change and Control in India

Benoy Krishna Hazra & Maitree Shee

8.1 INTRODUCTION

People need a better environment to live. For achieving a better environment people need to follow a proper balance between environment and developmental activities (being done within the environment). But if people go beyond their limits then the environment has to suffer and we being the survivors within this environment will suffer inevitably. As of now, the environmental impacts of climate change could be obviously seen in locales, a long way from where the larger part population live i.e. in the poles (the rapid melting of glaciers). This is just a wistful indicator of a major disaster in the near future. Presently only the farsighted researchers can see the climate change. Environment Scientists in the year 1995 conducted a World Environmental Conference in Reo De Janerio, and they emphasized on the effort to build consciousness among people. But the message was neglected time and again. The total of open states of mind or convictions offers basic conversation starters about the adequacy of almost a quarter century government funded training, decree, and engagement approaches which finish a complex scientific issue critical and fundamental for the intended interest group. The new compositions on open states of mind and alternate parts of the correspondence procedure are as yet being delivered to add to it and to comprehend the complaints and pleasantries for more compelling acquisition to environmental change correspondence. In spite of the fact that we are additionally frightened about the absence of individuals' support, we don't concur with the superficial analyze that individuals obviously 'couldn't care less' or 'are unmindful'. Rather, we assume critical bits of knowledge can be chalked out from the way environmental change is conveyed till date and how individuals have translated it. One might say that correspondence of environmental change is a testing one; it's an 'immaculate tempest'.

This test has restricted its part in empowering open engagement and support for activity.

8.2 ENVIRONMENT AND DEVELOPMENT LINKAGE

The discussion above made it apparent that the need to move from the non-literal thought of improvement has been long past due. The world today is at the cross roads of technological revolution and environmental emergency. Now the innovative insurgency stays as the support of human improvement and the idea has changed drastically with an expanding accentuation on a situation comprehensive advancement worldview. The current appraisal report of the intergovernmental board on environmental change has discovered our feelings of trepidation that condition as an idea indistinguishably connected with improvement has, by and now, stayed disregarded over the globe. The Brundtland Report underscored in unambiguous words: Failures to deal with the earth and to maintain improvement debilitate to overpower all nations (Report on the World Commission on Environment and Development: Our Common Future, 1987). Environment and development are unyieldingly connected. Advancement can't outlast upon a breaking down ecological asset base; similarly, the earth can't be protected when development causes to stop. These issues cannot be dealt with independently by divided establishments and arrangements; rather a complex system of ‚cause and effect' is conjoined with it. Since we are discussing an integrated phenomenon, the conceivable methodologies towards tending to the same too need to possibly be coordinated. Thus, integrated problem-solving methods are to be developed to solve the problems concerning an environmentinclusive development paradigm from the root itself (Mitra, 2014).The essential point here is that, since human is habituated to advancement, the fundamental approach towards improvement must be identified with human and are likely to act as the agents of the desired change. However, a huge question emerges here - by what means can person are locked in towards effectively adding to the constructive development of the foundation of an environmentinclusive development paradigm. In this context, the media can become a potential medium of reaching out to the masses and formulate them to receive desired public opinion from them. For this, the media needs to be two-dimensional. There is a need of feeling of environment consciousness among the media professionals (Bhattacharyya, pp. , 105-118). Environment concerns a vital social issue in the present situation, and the media can assume an essential part in tending to the same, it is normal that we begin truly considering a change in the existing media education system, from being an environmentexclusive exercise to an Environment Inclusive Media Educational System (EIMES)(Mitra, 2014).

8.3 COMMUNICATION, ENVIRONMENTAL & DEVELOPMENT PARADIGM

Primarily media was brought into the society to inform the common masses about the various decisions and steps taken by the authority in power but later with the advent of the ages media took the initiative to educate, and entertain the masses, as well as to develop an eco-conscious among them. Gradually the media professionals were equated with people who should follow couple of fundamental obligations and duties in an environmentinclusive development paradigm:

1. Sharing *facts and information with the masses*
2. *Spreading awareness among people*
3. *Integrating and mobilizing the masses*
4. *Analyzing, discussing and formulating public opinion "Give light and the people will find their own way." (Bhattacharyya, 2013)*
5. *Respecting environmental ethics*
6. *It is meant to serve people*
7. *Respecting the 7C's of communication: Credibility, Context, Content, Clarity, Continuity and Consistency, Channels, Capability of the audience.*

Environment' as the name suggests was thought to be a topic which could be well dealt only by the „environmentalists,' but this concept has changed from the moment media was used for expanding awareness about the environment. Environment inclusive media education framework started to raise an appropriate comprehension of the idea and extent of ecological training. Environmental training is the way toward perceiving values and clearing up ideas to create abilities and states of mind important to comprehend and welcome the interrelatedness among man, his way of life, and his biophysical environment. Ecological instruction additionally involves rehearse in basic leadership and self-planning an implicit rules about issues concerning natural quality (Division of Science, 1985).The principal Intergovernmental Conference on "Environmental Education" was sorted out by UNESCO in relationship with UNEP at Tbilisi, U.S; the Union of Soviet Socialist Republics, in 1977 worried on the significance of natural instruction in keeping up and enhancing the worldwide condition. It reinforces the primary destinations of enviornmental instruction which ought to create mindfulness and sympathy toward the total populace about the biological, monetary, political and social cooperation and common reliance in the earth and its issues (Division of Science, 1985). This might cause a major change in the climate. The conference established some objectives for environmental education which would build up the accompanying qualities in people and social gatherings:

a) awareness towards environment and its problems;

b) basic information about human and nature relationship;

c) social values and attitudes which are in concord with the environmental quality;

d) the skills to solve environmental problems;

e) the ability to assess environmental measures and education programs; and

f) Growing conscientiousness and urgency towards the environment in order to guarantee proper activities to tackle ecological issues.

The conference at Tbilisi confirmed that "environmental education" was not a new topic to discuss, but it shows a new facet in the existing curriculum subsumed with different discipline. Thus, we find that the environment as an issue of concern is not restricted to environmentalists only. It rather includes all people and associations who are indispensably related with it. The media and the media experts are no special cases to the same. The UNESCO Report entitled „Media as Partners in Education for Sustainable Development', asserted the fact that, media can assume a fundamental part in the era of biological cognizance among the common masses(Mitra, 2014).The conference naturally inspires a call for the need of formulating and implementing the Environment Inclusive Media Education System (EIMES) which shall concentrate on the acquirement of the following objectives by the media professionals which are indistinguishably associated with the necessary environmental education objectives:

a) Awareness (total environment and its related problems);

b) Knowledge (a basic understanding of, the environment and its associated tribulations);

c) Attitude(the motivation for actively participating in environmental improvement and protection);

d) Skill (to identify and solve environmental problems);

e) Partaking (prospect to be actively caught up at all levels of environmental problems).

From the study till now we can derive the idea that it is very important to teach the media personnel so that they could spread the idea to the masses with intelligence. This collective endeavor of the environmentalist as well as the media personnel could help to reach the goal of reduced outcome of climate change on the environment and promotion of sustainable development which is so vital to the reason for an environmentinclusive development paradigm.

8.4 CLIMATE CHANGE AND COMMUNICATION

Communication" which is perceptive of the applicable writing offers a more advantageous possibility of accomplishing the target of building a successful and reminiscent open engagement. We affirm that discussion on environmental change has been less helpful than one may try. The four primary reasons that the communicators like S.C. Moser and L. Dilling have accepted are:

a) An absence of data and understanding clarifies the absence of support of the general population, and in this manner, require more data and clarification to move individuals to activity. („Inspiring people with proper information');
b) Dread and dreams of potential calamity therefore of inaction would engage gatherings of people to activity („Motivating by fear');
c) The scientific planning of the subject would be most convincing and pertinent in moving common masses to activity („One size fits all'); and
d) Mass communication is the best approach to contact groups of onlookers on this matter „Mobilization through mass media' (Moser S. C., pp. , 161-174)

Inspiring and informing people about Climate change is a challenging issue because it's principal culprits i.e. carbon dioxide and other heat-trapping greenhouse gasses are colorless and odorless, and the long-term changes have emerged into notice only recently. The problem is building so slowly that it is beyond the noticed of the business entrepreneurs and others as they all are ignorant about the symptoms. For such a group Communication is essential because lack of knowledge about the details of climate change is NOT what averts more concern and activity. While acquaintance about the causes of climate change has connection with appropriate behavioral changes and proper understanding of system's prime audiences, there are also instances where better acquaintance about climate change does not essentiallyhoistalarm.

The climate is changing as a result of environment pollution, and this pollution can be natural, or human-made. Use of fossil fuel and carbon emission are major causes of the climate change. Both developed nations and developing nations are equally responsible for the climatic changes because of the following major factors:

a) Industrialization;
b) Urbanization;
c) More consumption of fossil fuel;
d) Greed of the human community;

e) Disappearance of natural negative caves by reckless felling of trees;
f) Not to adore to the suggestions of environmental scientists;
g) Not to follow the outcome of the Earth's Summit, wraths political decision becomes more important than scientists decision, and as a result greenhousegasses are increasing to the unwanted level;
h) Eco-friendly technology is not followed all over the world;
i) Unawareness among the people.

Though these are the harms done by humans; thereis some reason behind the climate change caused by nature. Natural causes like the eruption of volcanoes; solar activity contributes little to the process of climate change. But keeping aside these natural causes, anthropogenic activity and their attitude towards nature is the primary reason for climate change. The paper *"Adapting to Climate Change in Pacific Island Countries: the Problem of Uncertainty"* by Jon Barnett, explores the issue of logical vulnerability and the way it hinders getting ready for environmental change and quickened ocean level ascent (CC and ASLR) in Pacific Island Countries. The paper starts by examining the issues CC and ASLR postures for Pacific Island Countries, and it investigates the constraints of the prevailing way to deal with powerlessness and adjustment.

The book on "Italian Fourth National Communication under the United Nations Framework Convention on Climate Change" reads about domestic and universal moves Italy is making to meet its dedication underneath UN Framework Convention on Climate Change and its Kyoto Protocol. The Fourth Assessment Report of the UN Intergovernmental Panel on Climate Change affirm the evidence of the off-putting effect of environmental change and shows that such effects are progressively representing a genuine hazard to the biological community, nourishment creation, the accomplishment of supportable improvement and the Millennium Development Goals and additionally to individual's wellbeing and security. Be that as it may, after an extreme study on situations, information gathering, and investigation re-began in the second 50% of 2006. We are presently in the course to audit the National Plan of GHG discharge Reduction, CIPE 123/2002 Act. The demonstration is attempting to refresh Italy's arrangements and measures to meet Kyoto Protocol objective for the principal spending period 2008 – 2012, with the usage of local strategies and measures for no less than 80% of the decrease exertion and in the utilization of the Kyoto systems up to 20%. Be that as it may, the environmental change challenge requires additionally long haul worldwide endeavors.

The current reviews have concentrated on the likelihood that environmental change will decline treat multiplier issues in parts of the world that as of now experience the ill effects of political and financial insecurity. The powerful component of the U.S. national security foundation has generally recognized this practice. Albeit late reviews have measured operational issues and conceivable answers for U.S. constrains abroad – mostly identified with strategic necessities, for example, fuel supplies, they have not straight forwardly tended to dangers to U.S. government space operations or the relief and adjustment choices open to the space part. This paper abridges the applicable discoveries of reports on the national security ramifications of environmental change and looks to energize talk on the parts and necessities of the space group (Vedda, pp. , 14-17). This paper says about the climate change and importance of the space community system and seeks an action plan. It reviews the recent study on climate change and has done an experimental study of the U.S. history of climate change. As my area of the research paper is on climate change this paper helps me to think about the role of the Indian Government. Therefore I include the laws and policies (objectives and principals) implemented by the Indian Government.

The paper "Climate Change: Hope, Despair, and Planning" says, Climate Change is a hypothesis not an observation. The observation: Climate change is overwhelming the students. The purpose of this paper: How to engage business entrepreneurs in thinking about climate change as a planning matter without glossing over or disregarding their growing cynicism about their futures? What I've noticed is that climate change is turning the previously hopeful into the newly despairing. That is, the sense that the future holds, the promise is being interrupted by a new sense that there might not be a future at all (Seltzer, 2012). Climatic changes and its teachings, and planning, taken by his Government, relates to Indian planning in my paper.

Climate change has lucratively represented a safety measures at least to such an extent that it is decisively establishing the political schema, and that there is an extensively accepted exigency attached to it, even though the implementation of tangible policies is uncertain and has not taken place universally. In this paper, we question the monolithic orthodoxy of relating climate change to security, which often has concentrated too much on traditional security conceptions. We draw attention to the very different constructions of climate change as a security threat, as they bear dissimilar implications for the policy debate.

The consequence of climate change is disastrous as the incident happened in Uttaranchal, a state of India in 2016. Climate changes will bring natural disaster in our society and also affect the civilization as well as the sustainability of flora and fauna. Climate change will impact our agriculture which is the backbone of our economy. Talking of sustainable use of natural resources and environmental protection Climatic change has direct influence on the renewable resources of: (a) natural vegetation or forest area of a country; (b) Oceanic resources like phytoplankton, zooplankton and fish and others sea animals natural negation plays final role in land ecosystem. The oceanic living beings maintain a food chain.Both these renewable resources help a country directly for sustainable economic growth of forest and marine resource-based industries.Forests play vital role in the land and ocean water in the water bodies of the earth to balance the Greenhouse Gases. When rapid industrialization and rapid urbanization took place by converting forest area and agricultural land into the advance settlement, the environment was hampering and ecosystem was destroying, due to the emission of carbon at a higher rate which is a threat for the biosphere of this earth system. The modern Industrial policy is being adopted to maintain the GHG balance by keeping reserved forests, plantation in the periphery and new industrial zone. Environmental Protection has become a mandatory goal as mentioned in the different Protocol of earth summits in Kyoto, Geneva,etc. since years.

8.5 GOALS & STARATEGY FOR ENVIRONMENT

Goal 1:

Deliver collectively as one organization with people from various service area that has shown exemplary innovation and progress in the direction of growing and implementing climate change communiqué and propensity stratagem with a couple of audiences. Awareness on the efforts and classes discovered from those stations in the approach that facilitate and inspire emulation via different gadgets. Precise emphasis is given upon that group in every vicinity that has scientifically demonstrated and illustrated the consequences of, and adoption to climate change.

Goal 2:

Pick the high-quality respondents to serve as "Climate Change Ambassadors" to effectively engage and encourage traffic, neighborhood groups and school structures to take private and collective mitigation and version movements. The effort doesn't aim in the direction of education employees to come to be climate change scientists however to put together experienced ears (spectators/listeners) with sufficient records to make knowledgeable choices about their personal actions.

Goal 3:

Ceaselessly supplant climate change communication and engagement activities encouraged in this document to all different applicable listeners as an act for retaining the Future Communications efforts.

This strategy of engaging people with Communication is named as "Communication and Engagement Strategy." This strategy will focus on providing information necessary for effective communication and engage the masses both internally and externally. Within this strategy, audience is recognized as either primary or secondary. Primary audiences are those who lead by example "Climate Change Ambassadors". Secondary audiences are folks who receive messages from the primary audiences and might be propelled to take some alleviating and/or adaptation movement.

8.6 CLIMATE CHANGE POLICIES AND NEED OF COMMUNICATION

This paper does not advance a specific environmental change strategies or supporter for specific approaches here; we point towards introducing the bits of knowledge from the multidisciplinary inquire about writing on how correspondence can be plan-completely composed and did so that people(from all segments of the general public) may unyieldingly connect with themselves in the assignment of environmental change.

In June 2008, Prime Minister, Dr. M. Singh discharged India's first National Action Plan on Climate Change laying out existing and future approaches and projects tending to atmosphere moderation and adjustment. The arrangement distinguishes eight center "national missions" going through 2017 and guides administrations to submit point, by direct utilization plans toward the Prime Minister's Council on Climate Change by December 2008. (TERI, 2008)

On June 30, 2008 India released its National Action Plan on Climate Change (NAPCC) that extends till2017. Its goal changed into gaining a sustainable improvement that enhances each economic and environmental aspiration and to allow India to turn out to be a rich and efficient financial system that is self-sustaining for both present and future generations. The NAPCC is ruled by a list of concepts; the most applicable of them are listed under:

First of all, setting up a sustainable advancement methodology that diminishes destitution, powerlessness and anthropogenic effect on environmental change;

Secondly, holding India's money related increment through a move in influence hones that supports manageability by bringing down the discharge of GHG; and

Finally, it the set up request of late markets, administrative and intentional instruments to fortify sustainable advancement.

The NAPCC's main objective is to achieve a sustainable development. It complements each economic and environmental aspiration and to permit India to become a wealthy and green economic system that is self-maintaining via the eight National Missions. It acts as an ambassador of a multifaceted, long-time period and unified approach to permit India to grow and to achieve the important objectives in the structure of weather trade. These objectives are operated by means of national governments, non-governmental agencies, the non-public regions well as other stakeholders. The climate change policies/ National Missions are:

a) First is,the National Solar Mission, which exploits India's sun based era limit by methods for offering to make sun oriented power forceful with customary sources. Assignments that advance sun based power incorporate the formation of a sunlight based research association; enhanced interest in overall anticipates innovative extension, fortifying local cooperation and creation of supporting solar powered power through government financing and ventures.

b) Second is, the National Mission for Enhanced Energy Efficiency, which advances a decrease in vitality admission focused nearer to huge businesses by means of permitting organizations to change control sparing authentications, providing charge diminishment motivating forces on vitality productive stock, and financing to diminish control utilization in both the private and non-private circles through different control applications. The NAPCC intends to infer 15% of India's vitality from sustainable sources with the guide of 2020.

c) Third is, the National Mission on Sustainable Habitat, which appears to expand vitality execution through city arranging, enhancing development codes, advancing the utilization of productive engines and moving inhabitants to utilize open transportation through financial motivations, and focusing on the significance of waste administration and reusing.

d) Fourth is the National Water Mission, which points towards the development of water utilize execution by 20% through putting new costs and opportunity approach that address the trouble of water lack and climate change.

e) Fifth is, the National Mission for Sustaining the Himalayan Ecosystem, which objectives to protect the wellbeing of both hilly and lush range biological communities in the meantime as also subsiding dissolving of the ice sheets.

f) Sixth is, he National Mission for a Green India, with the goal of expanding their forest the target of expanding their forest region from 23% to 33% and reforesting six million hectares of corrupted lush land region.

g) Seventh is, the National Mission for Sustainable Agriculture, with the goal of expanding the target of expanding the strength in their rural zone by method for putting resources into climate flexible jeans, expanding atmosphere scope of putting resources into climate versatile plants, expanding atmosphere scope frameworks and enhancing rural procedures.

h) Finally is, the National Mission on Strategic Knowledge for Climate Change, which intends to extend their insight into environmental change science, inquire about snags, demonstrating techniques, worldwide participation and setting up a Climate Science Research Fund. It advances private division activities and engagement in developing adaptation and moderation innovation by the utilization of capital assets.

The NAPCC comprises of other continuous targets which incorporate ceasing wasteful coal-terminated plants while making an interest in re-production quality from inexhaustible resources accomplish a positive rate of the aggregate lattice's quality admission under the Electricity Act of 2003 and National Tariff Policy of 2006. It likewise advances control proficiency under the Energy Conservation Act of 2001 by targeting enormous ventures through quality reviews and vitality marking bundles (Gough, 2016).

Climate pledges around the world

Table 8.1: This information appeared in an article named 'India unveils climate change plan,' published in The Guardian on 2/10/2015.

SL.No	Country	Pledge
1)	EU	Promised to cut emissions 40% by 2030 against 1990 level
2)	US	Cut by 26-28% by 2025 against 2005 level
3)	India	Cut off up to 25% by 2020
4)	China	Will top discharges by 2030 and diminish carbon intensity 60-65% by 2030 against 2005 level
5)	Australia	Pledged to cut emissions 26%-28% of 2005 levels by 2030
6)	Japan	Will cut greenhouse gas emissions 26% from 2013 levels by 2030
7)	Indonesia	Said it would cut emissions 29% by 2030 compared to what it is currently on course for ('business-as-usual)
8)	Brazil	Intends to cut emissions 37% by 2025 from 2005 levels
9)	Mexico	Pledged to make emissions 22% lower by 2030 than business-as-usual
10)	Kenya	Will cut emissions 30% by 2030 below business-as-usual
11)	Canada	Focused on decreasing ozone depleting substance emissions 30% below 2005 levels by 2030

The climate change relief and adaptation actions can augment the hazard of struggle, in addition to compound vulnerabilities in positive areas. The proceedings are primarily based on studying the evidence that violent political struggles arise over the dispersion of advantages from herbal resources (Watts, p. IPCC WGII AR5). Therefore, in instances wherein property rights and peace making organizations are non-working or illegal, efforts to conform climate change that modifies the distribution of get admission to resources **that** have the potential to create and aggravate conflict.

Maladaptation or greenhouse fuel mitigation efforts at odds with nearby priorities and assets rights may additionally increase the hazard of struggle in populations, specifically where establishments governing get right of entry to assets are vulnerable, or prefer one institution over some other (O'Neill, pp. 211-213; Butler, pp. 49(1)23-34; McEvoy, pp. 22(2), 353-363). Research at the speedy growth of bio-fuels production consists of studies associating land grabbing, land dispossession, and social clash (Molony, pp. 109(436), 489-498; Borras Jr, pp. 37(4), 575-592; Dauvergne, pp. 37(4), 631-660.; Vermeulen, 2010, pp. 37(4), 899-916). One study has identified possible links between elevated bio-fuel's production, food price spikes, and social instability consisting of riots (Johnstone, pp. 11-17). Eriksen and Lind in 2009discovered that climate change adaptation interventions in Kenya have disturbed surrounding conflicts. Different reviews focuses broadened utilization of atomic power, expanding the possibility of atomic multiplication or incidents of nuclear terrorism (Socolow, pp. 31-44; Steinbruner, 2012).

Ongoing armed struggle threatens many activitieswhich might be required to conform to climate change. Armed warfare crumbles up advertises and decimates infrastructure, limits education and the improvement of human capital gives way to loss of life and harm to workers and decrease the potential of the state to secure credit. The battle subsequently makes neediness and constraints livelihoods: that during turn will increase vulnerability to the impacts of climate change (Nigel, pp. 23-34; Deng, pp. 231-250; Bockstael, pp. 1042-1053).

Climate change has the capacity to increase competition between international locations over shared resources. As an instance, there is challenge about contention over changing resources inside the Arctic and trans boundary river basins. Climate changes constitute a mission of the effectiveness of the various establishments that manage relations over those resources, but, there is high scientific agreement that this improved competition is not going to guide directly to struggle among states. The evidence so far suggests that the character of resources which include Trans boundary water and set up of conflict decision establishments have been able to avert rivalries in ways that keep away from violent conflict. (W. Neil Adger, 2013)

8.1 POSSIBLE SOLUTIONS & ALTERNATIVES

In a country where neither alarmism nor Pollyannaism seemed to yield preferred results, smart incorporation of techniques can also drive result in more noteworthy engagement. Firstly, conversation that avows rather than threats the sense of self basic world-views held with the aid of target market has been proven to create more openness to risk information's.(Barman, pp. 45(1): 1–65)Secondly, facts about threat and images that evoke worry should be strained and usually combines with messages and facts that offer particular, even minded help in acknowledging doable solutions. At long last, given the ideological polarizations around reactions to climate change (discussed in the next section of the chapter), the experience of fear and being overwhelmed, and the deep and enduring societal modifications required to address the problem. There may be an important place for facilitated dialogue and organized pondering of the issues as they develop(Braman, pp. 45(1): 1–65).

In this chapter, we're enunciating the role of communication that is broader than a few communicators would possibly assume: it isn't just a manner of conveying facts or prompting an audience or a passive receiver for persuasion. As a substitute, we advise that humans and the entrepreneurs in a democratic society are best served through actively engaging with an issue, making their voices and values heard, and contributing to the components of societal responses. Enforcing a deluge of scientific records and technocratic solutions on a populace without discussion and focus of risk and

Communication for Climate Change and Control in India

choices is possibly to lead to resistance and competition.(Moser S. C., pp. 161-174)

The term „engagement' illustrates three archetype assumptions on which our chapter rests. Firstly, science alone can never compel us to action as itdepends on a value-driven interpretative judgment. We believe that science has already established that climate change is in an operative stage and that the problem is mostly human-made, and it is causing significant threats to human and its environmental systems. This study sets us to take immediateand significant actions to reduce climate change-related risks. Secondly, in a democracy, any big action requires public input and support and its implementation requires the active engagement of the public. And thirdly, „communication' which is an essential means to link entrepreneurs, politicians, and the public, should play a constructive role in enabling public engagement. Here we could well utilize the Three-point Communication Model.

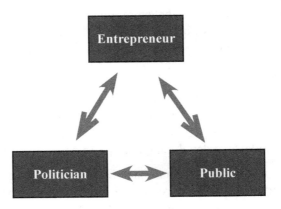

Fig 8.1: Three point Communication Model

The three point communication model has three focal points i.e. Entrepreneur, Politician and the Public. All the three factors vary from each other in various ways. The Entrepreneur is the business planner and for them making aprofit is the major goal. They would design their business for the people's demand and would try to produce as much as possible to meet the increasing demand for the product without keeping the matter of environmental effect into their considerations. In such a circumstance we the advocates for environment protection should come up with such a communication plan that would produce an interest among the entrepreneurs, and they would readily accept it. The language of our communication plan should be more appealing for the entrepreneurs, and they should feel a push in them to re-think in the way we want them to think. On the other hand, our communication plan should cater to both the entrepreneur's interest as well as the protective endeavor of the climate conservationists. The Politician is the person or an organization that that is entitled to look after various activities in its field of work, who gives permission to any developmental construction. When the term „development' is involved, then it should truly be so. Hence, when a climate conservationist is speaking to a politician or a political organization he should design the language lawfully, that would interest the organization. The communication plan should point out the advantages and disadvantages of any developmental work and should draw a road map for the political organization, to give them an idea of which direction to go. This roadmap should consist of the demands of commoners, the interest of the entrepreneur without violating any environmental law mentioned in the Book of Constitution, of India. The public is that section of the triangle that could be molded into the desired shape, but they are the real sufferers or gainers of any kind of developmental project. The communication plan for this section is the most sensitive that would help in gaining their faith. The people should be convinced in such a way that they might create pressure on the entrepreneurs as well as the politicians to maintain the right method of the developmental program.

Hence, it could be said that the level of language and the communication plan for the three broad sections of the society needs three different styles of communication but the goal should be same. The arrows in fig.1 indicate the flow of two-way communication even between the three major heads of the society. Thus it is clear from the figure that communication between all the elements is smooth and regular that would definitely reduce the chances of miss-communication and forming miss-conception. Communication is the biggest tool that could efficiently and effectively reduce any kind of barrier among different sections. This could be like a delta of information adding development in the concerned society.

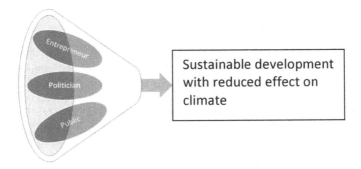

Fig. 8.2: Different Communication Plan Leading Towards the Same Goal of Reduced Effect on Climate

This path must additionally recollect the realities of current budget climate. Therefore, weather change engagement and verbal engagement strategies which might be numerous in method and now not expected to be universally accepted are in all likelihood to be more successful. Moreover, those strategies have to, in maximum times, be budget-impartial. "Budget neutral" does not equate to "low priority," but as a substitute requires the pursuit of activities that place a concern on innovation, capitalization of existing communications and outreach resources, leveraging with key partners and maximizing creativity (Climate Change Communications and Engagement Strategy for the National Wildlife Refuge System, 2014 February).

This Communication and Engagement Strategy approach report additionally seeks to expand critically-needed support from various sectors of our public for service efforts to mitigate and adapt to climate change on all refuges. Primarily, this strategy aims to improve the Refuge System's ability to engage and elicit tremendous modifications in adaptation and mitigation conduct amongst target audiences.

The goals of economic and social development need to be defined in phrases of sustainability in all international nations - developed or developing, market-orientated or centrally planned. Interpretations will vary, but should share certain general features and ought to go with the flow flow from a consensus on the primary concept of sustainable development and on a huge strategic framework for acquiring it.

8.9 CONCLUSION

This paper fundamentally concentrates on looking for an 'accord on the essential idea of lessening the inclination of climate change' and formulation of a strategic framework for achieving it.

The forthcoming environmental crisis that poses a potential threat on the global stage inexorably threatens to wipe out the human civilization in the days to come. The matter needs an immediate check. India, being an emerging superpower on the world stage, cannot avoid the inevitable crisis. This, however, appears to be probably not going to happen without a dynamic and mindful media contribution in guaranteeing individuals' cooperation and in viably tending to the worries of natural emergency. The media, being an influential medium of reaching out to the masses and in the formulating of desired public opinion, cansprout out as a catalyst towards the formation of an environmentinclusive development paradigm. Notwithstanding, the coveted media association can't in any way, shape or form be achieved without the ponder cooperation of the media experts be that as it may, just having the will to take part is not going to lash until Media experts have the expert intuition to play out their obligations and duties on the ecological front in the coveted way.

This necessarily calls for a vibrant Environment Inclusive Media Education System (EIMES) for all the sections i.e. Entrepreneur, Politician, and Public, which shall focus on infusing the sense of responsible media functioning towards environmental awareness. Once we are able to achieve that, we are probably going to get one bit nearer towards handling the environmental crisis in today's scenario. The desire to appreciate the vision of an environmentinclusive, sustainable development paradigm in the new millennium both in the Indian context and the world at large would be fulfilled only with the wake of consensus on the basic concept.

BIBLIOGRAPHY

1. Barnette, J. (2001). "World Development"Adapting To Climate Change In Pacific Island Countries: The Problem Of Uncertainty, 977-993.
2. Bhattacharyya, K. K. (2014) "Media Watch"Revisiting the Contours of Media Education: A Study in the Indian Context, 105-118.

3. Bhattacharyya, K. K. (2013) Science Communication As A Tool For Development.
4. Bockstael, H. a. (2011). "Journal of International Development" Diamond Mining, Rice Farming And A „Maggi Cube': A Viable Survival Strategy In Rural Liberia?, 23(8), 1042-1053.
5. Borras Jr, S. P. (2010). "Journal of Peasant Studies" The politics of biofuels, land and agrarian change: editors' introduction., 37(4), 575-592.
6. Braman, K. a. (2008). "American Criminal Law Review"The self-defensive cognition of self-defense, 45(1): 1–65.
7. Butler, C. a. (2012). "Journal of Peace Research" African range wars: Climate, conflict, and property rights, 49(1), 23-34.
8. "U.S. Fish and Wildlife Service (National Wildlife Refuge System)" (2014 February). Climate Change Communications and Engagement Strategy for the National Wildlife Refuge System.
9. Dauvergne, P. a. "Journal of Peasant Studies"(2010). Forests, food, and fuel in the tropics: the uneven social and ecological consequences of the emerging political economy of biofuels, 37(4), 631-660.
10. Deng, L. "African Affairs" (2010b). Social capital and civil war: The Dinka communities in Sudan's civil war., 109(435), 231-250.
11. Division of Science, T. a. (1985). Environmental Education: Module for Pre-Service Training of Social Science Teachers and Supervisors for Secondary Schools. Retrieved June 2014, from http://unesdoc.unesco.org: http://unesdoc.unesco.org/images/0006/000650/065036e.pdf
12. Mitra, A. and Bhattacharya, K. (2014) Environmental Crisis and Media Responsibility: A Study in the Indian Context. (n.d.).
13. Franziskus von Lucke, Z. W."8th Pan – European Conference on International Relations. Warsaw, Poland" (2013),What's at Stake in Securitising Climate Change? Towards a Differentiated Approach.
14. Gough, A. J. "Journal of Earth Science & Climatic Change"(2016) India's Energy-Climate Dilemma: The Pursuit for Renewable Energy Guided by Existing Climate Change Policies.
15. Jamieson, & Steffen. (2011) The oxford handbook of climate change and society.
16. Johnstone, S. a. "Survival"(2011)Global warming and the Arab Spring, 53(2), 11-17.
17. McCright, D. a., & Versteeg, H. a. (2011)The oxford handbook of climate change and society.
18. McEvoy, J. a. "Global Environmental Change"(2012). Discourse and desalination: Potential impacts of proposed climate change adaptation interventions in the Arizona–Sonora border region, 22(2), 353-363.
19. "Ministry for the Environment, Land and Sea." (2007). Fourth National Communication under the UN Framework Convention on Climate Change. Italy: Ministry for the Environment, Land and Sea.

20. Molony, T. a. "African Affairs"(2010)Biofuels, food security, and Africa, 109(436), 489-498.
21. Moser, S. C. "The oxford handbook of climate change and society"(2011) Communicating climate change: closing the science-action gap, (pp. 161-174). Oxford: Oxford University Press.
22. Nigel, J. "Norsk Geografisk Tidsskrift-Norwegian Journal of Geography "(2009)Livelihoods in a conflict setting, 63(1), 23-34.
23. Norgaard, K. M. (2011)The oxford handbook of climate change and society.
24. O'Neill, B. a. "Global Environmental Change"(2010). Maladaptation, 20(2), 211-213.
25. "United Nations"(1987) Report on the World Commission on Environment and Development: Our Common Future
26. Seltzer, E. "ACSP"(2012). Climate Change: Hope, Despair, and Planning.
27. Socolow, R. a. "Daedalus"(2009) Balancing risks: nuclear energy & climate change, 138(4), 31-44.
28. Steinbruner, J. S. (2012). Steinbruner, J., Stern, P. and Husbands, J. (eds.) "Committee on Assessing the Impact of Climate Change on Social and Political Stresses"Climate Change and Social Stress: Implications for Security Analysis, Washington D.C: National Research Council: Washington D.C.
29. Steve McIntosh, C. P. "Boulder, USA: The Institute for Cultural Evolution."(2012) Campaign Plan For Climate Change Amelioration.
30. TERI. (2008). CLIMATE CHANGE MITIGATION MEASURES IN INDIA. New Delhi.
31. United Nations. (1987). Report of the World Commission on Environment and Development: Our Common Future. Retrieved December 31, 2013, from http://conspect.nl: http://conspect.nl/pdf/Our_Common_Future-Brundtland_Report_1987.pdf
32. Vedda, J. A. "Pasadena, California: American Institute of Aeronautics and Astronautics"(2009). Climate Change Threats to National Security and the Implications for Space Systems, (pp. 14-17).
33. Vermeulen, S. a."Journal of Peasant Studies" (2010). Over the heads of local people: consultation, consent, and recompense in large-scale land deals for biofuels projects in Africa, 37(4), 899-916.
34. W. Neil Adger, J. P. (2013). Human Security.
35. Watts, P. "G. D. Jon Barnett (Australia)"(2001)Human Security, IPCC WGII AR5.

Chapter 9

Of Climate Change & The Calamitous Events of Kashmir Ecology (16th C. Onwards): An Historical Analysis

Mumtaz Ahmad Numani

9.1 INTRODUCTION

Of the world over, scientists/ geo-scientists/ social-scientists through their pioneering research have provided conceptual models that immensely demonstrate a direct cyclic relation between the climate, the landscape and calamitous events of any territorial unit of the planet earth. The first fact though mentioned by all the chroniclers about Kashmir was its invigorating climate. But, in our age of environmental crisis, the meaning and feeling of invigorating climate, goes irrelevant more often than not. Climate is a relative term the change of which is thus real. It has got an everlasting impact on the biotic and abiotic components of the environment. Interestingly, calamitous events in past are more attributed to fate than to climate change. Fate like a mad sovereign, Shrivara[1] rhetorically opines, can in a moment bestow unusual favour on his subjects when propitious, and inflict untold miseries when unpropitious. But who can understand the caprices of fate Shrivara asserts in his tone? Shrivara also informs us of an untimely snow fall that was followed by famine in Kashmir. Although climate failure in Kashmir: even occasionally had troubled people in the past, but its persistent failure above average is a matter of due concern in the

---

[1] Shrivara is the celebrated author of *Jainarajatarangini*, (The history of Kashmir), that adds to the previous works of Kalhana and Jonaraja, thereby providing an update of the history of Kashmir till 1486 CE.

present. Of several, but two important objectives may define the purpose of writing this paper. One, on the basis of sources is to recapitulate and offer a critical comprehensive analysis of the calamitous events occurring during **16th century onwards Kashmir**. Two, is to briefly examine the current and future challenges of the state in seeking wiser conceptual models for addressing the issue of climate-change and the calamitous events frequently occasioning in Kashmir.

The principal sources of knowledge of the history of Kashmir have mostly been documented in four languages i.e., Sanskrit, Persian, Urdu and English. The manuscripts, travelogues and reports help us in unravelling the historical events of the past to improve our understanding in present. The present study based on select principal sources put together may remain insufficient, but substantially significant.

As the main Mughal accounts (like: *Tarikh-i Rashidi, Ain-i Akbari, Tuzuk-i Jahangiri, Shahjahan Nama* and the other) have some passing references only but not detailed content recorded on the calamitous events of Kashmir ecology. Therefore, much information here put together is based on two important source books: *Waqiat-i-Kashmir* and *Tarikh-i-Hassan*. Both these texts are originally written in Persian and later on translated into Urdu. These contain rich information on the history of Kashmir ecology. Particularly in *Tarikh-i-Hassan*, calamitous events occasioned are discussed in detail and date-wise. But nonetheless, discrepancies in some cases do occur but negligible.

9.2 CALAMITOUS EVENTS

9.2.1 Earthquakes:

In Kashmir, writes Vigne, before 1828, there had been no earthquake within the memory of any living person except one that occurred about fifty years ago [1779][2], which was rather severe to last at

---

[2] Though Vigne doesn't provide us any detailed information, but, Hassan informs that, during the rule of Karim Dad Khan, Kashmir in 1779 witnessed an earthquake which in a twinkle of an eye damaged houses in the city and villages. Good number of people lost their life. About one and a half month severe

intervals for a week. However, he adds that, an earthquake is mentioned in Prinsep's tables as having taken place in 1552[3] (Figure1).The earthquake of 1828 occurred on 26th of June which

---

earthquake-jerks were felt continued. People had built makeshift accommodation in open fields. See, Pir Ghulam Hassan Khuihami, *Tarikh-i-Hassan*, Ed. Sahibzada Hassan Shah, (Srinagar, 1954) Vol. 1st. The Research and Publication Department, Srinagar, p. 470;(Here after, *Tarikh-i-Hassan*, Vol. 1st); Pir Ghulam Hassan, Khuihami, *Tarikh-i-Hassan*, Ed. Sahibzada Hassan Shah, (Srinagar, 1954), Urdu transl. by Shamsuddin Ahmad, *Shams-ut-Tawarikh*, (Srinagar, 2003), Vol. 1st. p. 453; (Hereafter, *Shams-ut-Tawarikh*). See also, R.N. Iyenger and et al., *Earthquake history of India in Medieval times*, in Indian Journal of History of Science, 34 (3), 1999, p. 192; (Hereafter,*Earthquake history of India in Medieval times*); Bashir Ahmad and et al., *Historical record of earthquakes in the Kashmir Valley*, in Himalayan Geology, Vol. 30 (1), 2009, p. 79; (Hereafter, *Historical record of earthquakes in the Kashmir Valley*).

[3] Khawaja Muhammad Azam Didamri and Hassan inform us that, during the rule of Ismail Shah, Kashmir in 1552 witnessed an earthquake which caused an unparalleled damage to the valley of Kashmir. It was felt as if the doomsday had arrived. It was only after two months, the earth got back its stability from small earthquake-jerks that occasionally continued for long. The impact was such that, several villages (For example, Hassanpur and Hussainpur) including major water bodies totally disappeared from the ground. (Moreover, prior to 1552 earthquake, Hassan also records the details of an earthquake that had occurred on 24th September 1501 A.D. when Fatah Shah was ruling the Valley). For details see, Khawaja Muhammad Azam Didamiri, *Waqiat-i- Kashmir*, Urdu Transl., Dr. Shamsuddin Ahmad, (Srinagar, 2001), pp. 142-143, (Hereafter, *Waqiat-i Kashmir*); *Tarikh-i-Hassan*, Vol. 1[st.] pp. 468-69; *Shams-ut-Tawarikh*, p. 452. However, here it bears to mention that, there is some difference of opinion about the earthquake-date recorded in historical accounts. For example, converting the Hijri date into years, Iyengar and et al., on the basis of some Sanskrit and Persian sources are of the opinion that, the earthquake that Hassan and others have described had taken place in 1552, had actually been taken place in 1555. See, *Earthquake history of India in Medieval times*. Whereas, Bashir Ahmad & et al., on the basis of some select sources have also ascribed the earth quake date to 1553 & 1555 respectively. See, *Historical record of earthquakes in the Kashmir Valley*. Since all the historical accounts describe the 1552/1553/1555 earthquake as the mega-earthquake of Kashmir. Interestingly, some geo-scientists on the basis of mapping &modern techniques have tried to find out the current locations and magnitude of the earthquake(see, Fig. 1),to put it straight on record. For details, see, Susan Hough and et al., *Kashmir Valley Megaearthquakes: Estimates of the magnitudes of past seismic events foretell a very shaky future for this pastoral Valley*, in, American Scientist, Vol. 97, No. 1 (Jan—Feb. 2009), PP. 42-49.

shook down some 1200 houses and 1000 persons were killed, perhaps.[4]

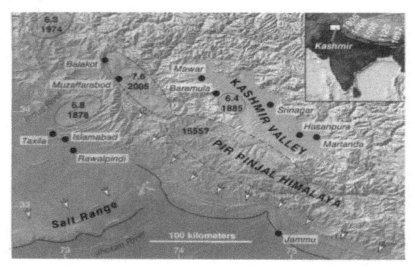

Fig. 9.1: Shows the locations and magnitudes of the mega-earthquakes of Kashmir. Susan Hough and et al., have tentatively assigned a surface magnitude value of 1552/1555 earthquake around 7.8/8.In the map, Susan & et al., highlight: "The dashed ellipses show a speculative location for a large 1555 earthquake. The inset shows the locations of great earthquakes along the Himalayan arc, showing that the Pir Panjal Himalaya region may be long overdue for such an event". Source: Adopted from, Susan Hough and et al., *Kashmir Valley Mega earthquakes: Estimates of the magnitudes of past seismic events foretell a very shaky future for this pastoral Valley,* in, American Scientist, Vol. 97, No. 1 (Jan—Feb. 2009), Fig. 2, p. 44.

---

[4] See, G. T. Vigne, Travels in Kashmir, Ladak, Iskardo: The Countries Adjoining the Mountain Course of the Indus, and the Himalaya, North of the Punjab, Vol. 1st London, 1842, pp. 281-282. (Here after, Travels in Kashmir, Ladak, Iskardo…,). However, according to Hassan, during the rule of Kirpa Ram, Kashmir on 5th of July 1828 witnessed an earthquake which caused huge damage. Lots of people lost their life. The effect was such that, severe earthquake-jerks were felt for three months and occasional aftershocks continued for nine months. See, Tarikh-i-Hassan, Vol. 1st p. 471; Shams-ut-Tawarikh, pp. 453-454.

Fires and flood, Lawrence[5] points out, however, sink into insignificance when compared with earthquakes, famines and cholera. Since the 15th Century eleven great earthquakes have occurred[6], all of long duration and accompanied by great loss of life. In the 19th century there have been four severe earthquakes, and it is worthy of notice that in the last two, 1864[7] and 1885, the most violent shocks were felt in an elliptical area whose focuses were Srinagar and Baramulla. The earthquake of 1885[8] commenced on May 30th, the shocks of which more or less were felt up to 16 August 1885. It is said that some 3,500 persons were killed, and the number of cattle, ponies and other domestic animals crushed by falling buildings was enormous.[9]

---

[5] Sir Walter Roper Lawrence (better known as Walter Lawrence) was an English author who also served in the Indian Civil Service under the British in India and wrote travelogues based on his experiences of traveling around the Indian subcontinent. He is the celebrated author of *The Valley of Kashmir* (1895) and *The India we Served* (1929).

[6] The details of which chronologically have been put on record by Hassan. See, *Tarikh-i-Hassan*, Vol. 1st pp. 468-473; *Shams-ut-Tawarikh*, pp. 451-455.

[7] However, Hassan records that the earthquake took placeduring the rule of Ranbir Singh in the year 1863. At most of the places in Kruhan and Baangal the earth tore into gaps but there was not much damage to life and property. Though the effect of earthquake continued for three months but itsintensity in comparison was felt lesser on the eastern side. See, *Tarikh-i-Hassan*, Vol. 1st. p. 471; *Shams-ut-Tawarik,*p. 454.

[8] Whereas, Hassan has put on record that: towards the closing years of Ranbir Sing's rule, on 30th May 1884 at 3:00am suddenly an earthquake took place—which caused people to utter huge hue and cry. The first jolt of earthquake leveled all houses in Sopore, Baramulla, Kruhan and Baangal. Also, in these areas much of floods caused ripping earth apart.At some places green coloured sand along with water happened gushing out. And the smell of Sulphur was felt for one month. For more details, see, *Tarikh-i-Hassan*, Vol. 1st pp. 471-73; *Shams-ut-Tawarikh,*pp. 454-55.

[9] See, Walter Lawrence, *The valley of Kashmir*, (London, 1895), 2nd ed. Srinagar, 2005, pp. 212-213,(Here after, *The Valley of Kashmir*); F. Bernier, *Travels in the Mogul empire*, transl. A. Constable, (revised by V. A. Smith), reprint: Delhi, 1983, pp. 394-395; (Here after, *Travels in the Mogul empire*). In foot notes, the translator of Bernier has only mentioned about the happening of previous centuries earthquakes (such as, 1552 and 1680—which according to Khawaja Azam Didamiri and Hassan has occurred inthe Subahdari of Ibrahim Khan in 1678, see, *Waqiat-i Kashmir*, p. 295; *Tarikh-i-Hassan*, pp. 469-70; *Shams-ut-*

The October 2005 earthquake **(Figure 1)**[10] centered near the city of Muzaffarabad and affected almost all parts of Kashmir. It is estimated that the death toll on both sides of Kashmir[11] could reach over 100,000. Approximately 138,000 were injured and over 3.5 million rendered homeless. According to government figures, 19,000 children died in the earthquake, most of them in widespread collapses of school buildings. The earthquake affected more than 500,000 families. In addition, approximately 250,000 farm animals died due to collapse of stone barns, and more than 500,000 large animals required immediate shelter from the harsh winter.[12] Here it bears to born in mind that, the worst affected major towns by 2005 earthquake on the Indian side were Tangadhar in district Kupwara and Uri in district Baramulla.[13]

Table 9.1: Earthquakes of Kashmir between 1550 and 1900 A.D.

Earthquake Date	Most Affected Area(s)	Summary
960 Hijri (1552	Hasanpora / Hussainpora/ Mawrah District	Described as mega-earthquake. Caused death of 600 people in Mawrah district of Kamraj Pargana alone. Some old villages/ springs /

Contd...

Tawarik, p. 453), but he escapes to mention any record on any of these earthquakes even in minor detail. Neither Vigne nor Lawrence put any substantial record about them. Interestingly, Aggarwal, too has become confused. For example, while mentioning about the earthquake details, he doesn't differentiate betweenthe earthquake of 1885 and the earthquake of 1680. Rather he mingles the details of one with the other. And thus confuses the earthquake of 1680 with the earthquake of 1885. See, *The Valley of Kashmir,* pp. 212-213; *Travels in the Mogul empire,* pp. 394-395; C M. Agrawal, *Natural Calamities and the Great Mughals,* 2nd Edition, Janaki Prakashan Patna: New Delhi, 1987, p. 83. (Here after, *Natural Calamities and the Great Mughals*).

[10] Susan Hough and et al., have assigned a surface magnitude value of 7.6 to the 2005 earthquake.

[11] Indian and Pakistan administered of Kashmir divide.

[12] See, G. M. Bhat, et. al., *Report of Centre for Disaster Studies and Research,* University of Jammu, Jammu (India), 2005, pp. 1-23.

[13] See, D. C. Rai& C. V. R. Murty, *Current Science,* Vol. 90, No. 8, 25 April 2006, pp. 1066.

		roads disappeared. Large damage to life and property happened.
1080 Hijri (1669 A.D.)	Not known	During the governorship of Saif Khan, an earthquake occurred. From dusk to dawn buildings shook like cradles. However there was not much damage to life.
(Between 1668/ 1678 A.D.)	Not known	During the governorship of Ibrahim Khan floods came—which were followed by earthquake. It caused much damage to property and life.
1148 Hijri (1736 A.D.)	Not known	During the deputy governorship of Dil Dileer Khan an earthquake occurred like a doomsday. In one jolt, it levelled the houses of the city and villages. The affect continued for three months. There was huge damage to life and property.
1148 Hijri (1736 A.D.)	Not known	During the deputy governorship of Dil Dileer Khan an earthquake occurred like a doomsday. In one jolt, it levelled the houses of the city and villages. The affect continued for three months. There was huge damage to life and property.
1193 Hijri (1779 A.D.)	Not known	In a twinkle of an eye, the earthquake, damaged houses in the city and villages. Jolts continued for one and half month. It took away many lives.
1199 Hijri (1784 A.D.)	Srinagar	It occurred during the governorship of Azad Khan. Its affect was much felt in the city where damage to life and property took place. The aftershocks continued for three months.

Contd...

1218 Hijri (1803 A.D.)	Srinagar	An immensely terrible earthquake occurred during the rule of Sardar Abdullah Khan. At many places houses collapsed. Women aborted and many people died due to the debris of walls.
1243 Hijri (1828 A.D.)	Srinagar	An earthquake occurred during Kirpa Ram's rule which shook down twelve hundred (1200) houses and killed nearly one thousand (1000) people. Severe earthquake jolts continued for three months.
1280 Hijri (1863 A.D.)	Kruhan and Baangil. (North Kashmir)	It occurred during Ranbir Singh's government. Although the aftershocks continued for three months but there wasn't much damage to life.
1302 Hijri (1884 A.D.)	Srinagar/ Sopore/ Baramulla	It occurred during the last days of Ranbir Singh's government. Land ruptured deeply at many places. It killed around three thousand five hundred (3,500) persons and huge number of animals.

Sources: *Tarikh-i-Hassan, Vol. 1st pp. 468-473; Shams-ut-Tawarikh, pp. 451-455; Waqiat-i- Kashmir, pp. 142-143, 295; Travels in the Mughal empire, pp. 394-395; Travels in Kashmir, Ladak, Iskardo..., pp. 281-282; The Valley of Kashmir, pp. 212-213.*

9.2.2 Floods:

In 1841[14], there was a serious flood, which Lawrence mentions, caused much debate to life and property. He says: "though I cannot

---

[14] Whereas, Hassan informs us that during the rule of Sheikh Ghlum Mouhi-uddin, in 1841 due to the break of Qazi-zada Band (embankment), Khanyar and Ranawari (town places in capital city of Srinagar) witnessed floods—which caused heavy damage to both of the places nearby. The continuous excessive rainfall for a week damaged six bridges between Fateh Kadal to Sambhal. See, *Tarikh-i-Hassan*, Vol. 1st. pp. 475-76; *Shams-ut-Tawarikh*, p. 457.

ascertain any accurate facts regarding the flood level of 1841, but some marks shown to me suggest that the flood of 1841 rose some nine feet higher on the Dal Lake than it rose in 1893. But the flood of 1893 was also a great calamity. It cost the state 64,804 rupees in land revenue alone, 25,426 acres under crops were submerged, 2,225 houses were wrecked and 329 cattle killed".[15]Lawrence also points out that, floods in Kashmir are caused by warm and continuous rains on the mountains which melt the snows or precipitate them down the hill sides into the streams.[16]However, due to the excessive rainfall, the state thereafter witnessed floods in the following years: 1903[17], 1905[18], 1909[19], 1929[20], 1948[21], 1950[22], 1957[23], 1959[24],

---

[15] *The valley of Kashmir*, pp. 205-206;Frederic Drew, *The Jummoo and Kashmir Territories: A Geographical Account* (1875), Reprint: Delhi, 1997, pp. 414-419; (Here after, *The Jummoo and Kashmir Territories...,*). According to Drew, the next havoc flood after 1841 occurred in the year 1858 that did serious damage at Naushahra--a place in Rajouri district of Jammu division.

[16] *The valley of Kashmir*, pp. 205-206. Moreover, here it bears to mention that, Hassan informs us on having taken place some more floods (not mentioned above) in Kashmir during the years: 1574-1575, 1668, 1730, 1735-1736, 1747, 1770, 1787-1788, 1799-1800 and 1836-1837 (see, table 2), the details of which have rarely been put on record by other modern historians. See, *Tarikh-i-Hassan*, Vol. 1st. pp.473-476; *Shams-ut-Tawarik*, pp. 455-457.

[17] Jarnail Singh Dev, *Natural Calamities in Jammu & Kashmir*, New Delhi, 1983, pp. 27-33. (Here after, *Natural Calamities in Jammu & Kashmir*). Here it bears to born in mind that, a report (cited below) states that, ten years after the flood of 1893, Srinagar witnessed another flood on 23 July 1903, which has been classified as the "greatest flood ever known", converting the city into "a whole lake". For details, see, a report titled, *"A Satellite Based Rapid Assessment on Floods in J&K- September 2014"*, Department of Ecology, Environment and Remote Sensing, Government of J&K, Bemina, Srinagar & National Remote Sensing Centre, Indian Space Research Organization, Hyderabad, (2014), p. 13. (Here after, *Report2014*).

[18] Natural Calamities in Jammu & Kashmir, pp. 33-36.

[19] Natural Calamities in Jammu & Kashmir, pp. 36-37.

[20] Report 2014, p. 13.

[21] Report 2014,p. 13.

[22] According to the *Report 2014*, it has been estimated that,nearly 100 people in the valley lost their lives due to the flood caused by the Jhelum's overflow. See, pages. 13-19.

1992[25] **(Figure 2, 3)**, 1996[26], 2006 **(Figure 4)**[27], 2010[28] and more recently in September 2014[29] **(Figure 5)** continuously. The overall devastation caused by these floods can hardly be counted.[30]

Table 9.2: Floods of Kashmir between 1550 and 1900 A.D.

Flood Date	Most Affected Area(s)	Summary
982 Hijri (1574 A.D.)	Not known	During Ali Khan Chack's government: floods came in the rainy season and destroyed houses and crops.
1089 Hijri (1668 A.D.)	Not known	It was Ibrahim Khan's period when floods occasioned and the houses like boats started floating around.
1143 Hijri (1730 A.D.)	Not known	During the governance of Nawazish Khan: floods occasioned due to the excessive rainfall in Kashmir. It damaged houses and crops.
1148 Hijri (1735 A.D.)	Not known	During the period of Dil Dalir Khan, floods occasioned in

Contd...

---

[23] Report 2014, pp. 13-14.

[24] Report 2014, p. 14.

[25] In terms of heaviest rainfall recorded since 1959, report 2014 states that, the 1992 floods were most devastating in terms of casualties. See, *Report2014*, p. 14.

[26] Report 2014, p. 14.

[27] *Report 2014*, pp. 14-15.

[28] According to the *Report 2014*, massive floods were caused by a cloudburst in the Leh-Ladakh region of J&K. For details see, page. 14.

[29] According to the *Report 2014*, on 4th September 2014, the state experienced 30hour long rainfall which broke the record of many decades in terms of loss it caused. See, page. 16.

[30] See, *Report 2014*, pp. 13-14.

			Kashmir—and destroyed thousands of houses. The abundance of water remained in the lawns for long.
1160 Hijri (1747 A.D.)		Jhelum and the periphery	During the governance of Afrasaiyab Khan, floods occasioned in Kashmir—and destroyed ten thousand (10,000) houses. The bridges of river Jhelum also collapsed.
1184 Hijri (1770 A.D.)		Jhelum and the periphery	The bridges of river Jhelum collapsed due to the floods occasioned during the governance of Amir Khan Jawan Shair. The buildings scattered hither and thither.
1202 Hijri (1787 A.D.)		Srinagar	During the time of Juma Khan, river Jhelum flooded and broke the Qazi-zada Band (embankment) due to which the northern part of the city submerged under water. Huge damage occurred to the houses and the other property of the inhabitants.
1214 Hijri (1799 A.D.)		Jhelum and the periphery	During the governance of Abdullah Khan, river Jhelum flooded and destroyed the crops outside in the fields.
1252 Hijri (1836 A.D.)		Bijbehara and Pampore	During the governorship of colonel Mian Singh, floods broke the bridges of Khanbal, Bijbhehara,

Contd...

		Amira Kadal and Panpore.
1257 Hijri (1841 A.D.)	Khanyar/ Rainawari	Floods occasioned in the city during the rule of Sheikh Ghlum Mouhi-uddin. The excessive rainfall damaged six bridges between Fateh Kadal to Sambhal.
1311 Hijri (1893 A.D.)	Srinagar	It cost the state 64, 804 rupees in land revenue alone; 2, 225 houses were wrecked and 329 cattle killed.

Sources: Tarikh-i-Hassan, Vol. 1st pp. 474-476; Shams-ut-Tawarikh, pp. 455-457; Travels in Kashmir, Ladak, Iskardo..., pp. 414-419; The Valley of Kashmir, pp. 205-206.

Fig. 9.2: Indicates floods in capital city Srinagar and adjoining areas in September 1992. Source: adopted from, *A Satellite Based Rapid Assessment on Floods in J&K- September 2014"*, has been prepared by the Department of Ecology, Environment and Remote Sensing, Government of J&K, Bemina, Srinagar & National Remote Sensing Centre, Indian Space Research Organization, Hyderabad, 2014, (Fig. 2), p. 14. (Here after, *Report 2014*).

Of Climate Change & The Calamitous Events of Kashmir Ecology... 209

Fig. 9.3: Indicates floods in Anantnag, Pulwama & Adjoining areas in September 1992. Source: adopted from, *Report 2014*, (Fig. 3), p. 15.

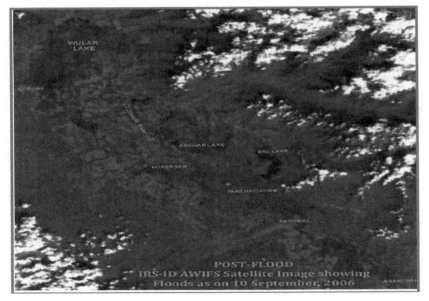

Fig. 4: Indicates floods in Kashmir valley in September 2006. Source: adopted from, *Report 2014*, (Fig. 4), p. 15.

210   Environment Impact Assessment

Fig. 9.5: Indicates floods in J&K in September 2014. Source, adopted from: *Report 2014*, cover page.

9.2.3 Epidemics:

A description of plague occurs in our sources, writes Habib, when it spread from Panjab to Kashmir, Delhi and Agra, and surrounding districts in 1615, and continued to rage till 1619, causing extensive mortality with practically none of the affected persons surviving.[31] Jahangir, as Habib points out, gives us an exceptionally careful description of its visitation. All accounts relate the spread of disease to rats and indeed, the bacterium *Pasteurella prestos* is known to be carried by fleas from infected rats. Jahangir noted that the epidemic followed a widespread scarcity, and was virulent in winter while becoming dormant in summer. But one wonders, rhetorically Habib asks: "whether the plague that Jahangir describes then travelled

---

[31] Irfan Habib, Man and Environment: The Ecological History of India, Tulika Books, New Delhi, 2011, P. 87, (Here after, Man and Environment: The Ecological History of India).

across the world to France to rage there in 1628-31, carrying off a million people".[32]

According to Vigne, in 1828 when the earthquake gradually ceased, the cholera made its appearance in the Valley. A census of the dead was taken at first, but discontinued when it was found that many thousands had died just in twenty-one days.[33] However, Lawrence opines that: "the first mention of cholera as far as I can ascertain was in 1598 A. D., and before that time the disease was either unknown or known by a name different to what is now used, Wa'ba"[34]. He continues to add that, in the 19th century, there have been ten epidemics of cholera[35], all more or less disastrous to the people of Kashmir, and of these, most probably the worst was the last which occurred in 1892 A.D. It is stated by the chief medical officer of Kashmir that 5, 781 persons died in Srinagar, and 5, 931 in the villages.[36] It is generally agreed, Lawrence points out, that the centre

---

[32] Jahangir, *Tuzuk-i-Jahangiri* (1624), ed. Syed Ahmad Khan, Ghazipur and Aligarh, 1863—64, Sir Syed Academy, 2007, pp. 219-220, 259-260. (Here after, *Tuzuk-i-Jahangiri*).The *Jahangirnama: Memoirs of Jahangir, Emperor of India*, transl., ed. and annotated by W. M. Thackston, New York, 1999, pp. 253-254, 291-292. (Here after, *The Jahangirnama....*); *Man and Environment: The Ecological History of India*, p. 87;*Natural Calamities and the Great Mughals*, pp. 54-55;*Tarikh-i-Hassan*, Vol. 1st. p. 466; *Shams-ut-Tawarikh*, pp. 449-450.

[33] *Travels in Kashmir, Ladak, Iskardo...*, p. 282. Whereas, Hassan informs that, the epidemics occurred during the period of Deewan Kirpa Ram in 1827. It took away the life of thousands of thousands people. See, *Tarikh-i-Hassan*, Vol. 1st. pp. 466-67; *Shams-ut-Tawarikh*, p. 450.

[34] *The valley of* Kashmir, pp. 218-219. Wa'ba in local Kashmiri dialect is a generic term used for epidemics.

[35] The details of which have been put on record by Hassan. Hassan informs us that during 1783-85, 1819, 1827, 1854, 1857, 1872, 1876 and 1878 (see, table 3), epidemics spread in Kashmir and caused huge damage to life. However, Hassan doesn't identify any specific reason for the spreading of epidemics in Kashmir during the years mentioned. The reasons, therefore, were might be various and different. For more details, see, *Tarikh-i-Hassan*, Vol. 1st. pp. 466-468; *Shams-ut-Tawarikh*, pp. 449-451.

[36] Lawrence elaborates to inform us that, he was in the camp during the epidemic and moved through some of the most infected Centers. He also believes that, owing to the panic which set in the registration in the districts was not as careful as it might have been in the city. Thus, the number 5,931 deaths does not seem him to represent the actual mortality from cholera in the villages.See, *The Valley*

and nursery of cholera in Kashmir is the foul and squalid capital Srinagar; but when it is once established there it soon spreads to the dirty towns and villages where no medical treatment was available.[37]

Here it won't go out of place to mention a modern scholar who also highlights some details of cholera in an essay entitled, "Recent History of Kashmir: An Account from the Issues of Environment".[38] In 1857-58, Pandita writes, cholera raged to December and January and then ceased, but reappeared after three months and prevailed for two months more. In 1872 an epidemic of cholera commenced from August and lasted four months. The epidemic took a severe toll in 1875-76 when it claimed the lives of at least 10,000 people. About the same number of people perished in a

---

of Kashmir, pp. 218-19; R. Harvey, *A Brief Sketch of the Epidemic of Cholera in Srinagar, Kashmir, May-June, 1892*, The British Medical Journal, Vol. 2, No. 1650 (Aug. 13, 1892), pp. 345-347.Here it bears to mention that, the actual reporting of mortality from 1892 cholerahas also been questioned by Ernest F. Neve. Neve records that, in thirteen years, Kashmir witnessed three invasions of Cholera. [In 1888, there were probably 18,000 attacks and a total mortality of nearly 10,000].In 1892 the official returns showed 16,845 cases with 11,712 deaths, owing to the concealment then extensively practised by the people, it is probable that there may have been hundreds of unregistered cases also. See, Ernest F. Neve, *A Brief Account of the Recent Epidemic of Cholera in Kashmir*, in, The British Medical Journal, Vol. 2, No. 2085 (Dec. 15, 1900), pp. 1705-1706. *(*Here after, *A Brief Account of the Recent Epidemic of Cholera in Kashmir*).Interestingly, in a report, W. Hill Climo (Brigade-surgeon-lieutenant-colonel) points out the dirty habits of the people as the main cause of the spreading of 1892 cholera in the city. See, W. H. Climo, *The Floods in Kashmir and Srinagar: A New Focus of Endemic Cholera*, in, The British Medical Journal, Vol. 2, No. 1702 (Aug. 12, 1893), pp. 396-397.

[37] Lawrence (and E. F. Neve) elaborates on the issue of the spread of cholera in Kashmir. He records that, it is a curious fact that the villages on the Karewa Plateaus seemed free from cholera and the disease was most rampant in the alluvial parts of the Valley. How the cholera arrives to the Valley is a matter of concern to be discussed by the competent authorities then. For example, before the road from Baramulla to the Punjab was opened, cholera might occur in India, while Kashmir was healthy, and whereas, there were twelve epidemics of cholera in the Punjab between 1867 and 1890, but there were only five out-breaks in Kashmir during the same period.See. *The Valley of Kashmir*, pp. 218-19; *A Brief Account of the Recent Epidemic of Cholera in Kashmir*, p. 1706.

[38] S. Bhatt, Kashmir Ecology and Environment: New Concerns and Strategies, New Delhi, pp. 117-128. (Here after, Kashmir Ecology and Environment: New Concerns and Strategies).

very virulent epidemic of cholera within two months in 1880.[39] And thereafter in1897 cholera prevailed for four months and killed thousands of people around.[1]

Another visitation of cholera, which commenced in August 1900, was introduced from the Punjab. Its origin in the valley, Pandita points out, was traced to the pilgrims for the Amarnath Cave.[40] Again an epidemic of plague prevailed from November 1903 to July 1904.[41] Thereafter, in the years of 1906, 1907[42], 1910, 1914, 1915 and 1918 cholera killed a huge junk of population in Kashmir. All the status show-case that, cholera like trade, travelled by roads in the valley.[43] Thus if Pandita is to be believed for numbers, the total number of persons killed by the spread of cholera during the years cited above sums up in thousands.

---

[39] Hassan informs that, the spread of cholera had taken place when Ranbeer Sing was ruling Kashmir. For details, see, *Tarikh-i-Hassan*, Vol. 1st. p. 467; *Shams-ut-Tawarikh*, pp. 450-451.

[39] *Kashmir Ecology and Environment: New Concerns and Strategies,* p. 123.

[40] Pandita points out that there were 19,265 cases and 10,811 deaths. Out of which 2,439 cases and 1,293 deaths were occurred at Srinagar only. See,*Kashmir Ecology and Environment: New Concerns and Strategies,* pp. 123-124. Here it bears to born in mind that, the question on the spreading of cholera from Punjab to the valley of Kashmir has also been raised by Lawrence and Ernest Neve. See, *The Valley of Kashmir,* pp. 218-19; *A Brief Account of the Recent Epidemic of Cholera in Kashmir,* p. 1706.

[41] According to Pandita, it killed 1,379 persons in various towns and 56 persons in Srinagar. See,*Kashmir Ecology and Environment: New Concerns and Strategies,* pp. 123-124.

[42] See, *India: Report from Calcutta—transactions of service—cholera, plague and smallpox—cholera epidemic in Kashmir,* in Public Health Reports (1896—1970), Vol. 22. No. 29 (July 19, 1907), p. 1005. Eakins (acting assistant surgeon) in June 1907 reports that: the cholera epidemic continues in Kashmir. The number of cases reported for the week (ended June 3) was 1,447 with 771 deaths. Since November, 1906, the total number of cases reported is 10,550, of which 5,689 were fatal.

[43] Kashmir Ecology and Environment: New Concerns and Strategies, pp. 123-124.

Table 9.3: Epidemics of Kashmir between 1550 and 1900 A.D.

Epidemic Date	Most Affected Area(s)	Summary
1024 Hijri (1615 A.D.)	Not known	It did spread from Punjab to Kashmir during the rule of Ahmad Baig Khan. The epidemic followed a widespread scarcity and thus took a heavy toll of life.
1198 Hijri (1783 A.D.)	Srinagar	The epidemic spread during the governance of Azad Khan. Thousands of people died in the city.
1235 Hijri (1819 A.D.)	Not known	It occasioned during the period of Deewan Moti Ram and took away the life of thousands of people.
1243 Hijri (1827 A.D.)	Not known	It occasioned during the period of Deewan Kirpa Ram and continued for one month. Many thousands died in days.
1264 Hijri (1845 A.D.)	Not known	It occasioned in the valley during the rule of Sheikh Ghulam Mohuiddin. Thousands of people were buried without wrapping even coffin on their bodies.
1274 Hijri (1857 A.D.)	Not known	It occasioned during the rule of Ranbir Singh in the months of Magh and Poh, and killed people in thousands.
1284 Hijri (1867 A.D.)	Not known	During the period of Ranbir Singh, epidemic spread for a period of three to four months. Thousands of people died.
1289 Hijri (1872 A.D.)	Not known	Again during the period of Ranbir Singh epidemic of cholera commenced from August and lasted for four months.
1293 Hijri (1876 A.D.)	Not known	Again during the period of Ranbir Singh epidemic

Contd...

		occasioned and claimed the lives of atleast ten thousand (10,000) people.
1296 Hijri (1878 A.D.)	Srinagar	Again during the period of Ranbir Singh epidemic took a heavy toll of life for fourty days.
1310 Hijri (1892 A.D.)	Srinagar	With its affect five thousand seven hundred eighty one (5, 781) persons died in Srinagar and five thousand nine hundred thirty one (5, 931) in the villages.
1318 Hijri (1900A.D.)	Srinagar	The epidemic of cholera spread in the valley in August 1900 through the pilgrims of Amarnath Cave from Punjab. One thousand two hundred ninety three (1, 293) deaths occurred at Srinagar alone.

Sources: *Tuzuk-i-Jahangiri*, pp. 219-220, 259-260; *Tarikh-i-Hassan*, Vol. 1st pp. 466-468; *Shams-ut-Tawarikh*, pp. 449-451; *Travels in Kashmir, Ladak, Iskardo...*, p. 282; *The Valley of Kashmir*, pp. 218-219; *Kashmir Ecology and Environment: New Concerns and Strategies*, pp. 117-128.

9.2.4 Famines

Habib is of the opinion that, with narrative histories coming to our aid, we can build a better famine and/ or epidemic record(s) for medieval and modern times than for the ancient history. Kalhana, as Habib points out, writing his great history of Kashmir the Rajatarangini in c. 1150 records a frightful famine in that valley occurring in 917-18, followed by one in 1099-1100 (caused by inundations) and another in 1123-24 (through internal war and carnage). In all the three famines, Kalhana says, human corpses dumped into the streams obstructed their courses. In 1596 extensive famine conditions prevailed owing to drought in northern India, and no relief was obtained the next year, during which much misery from famine was reported also from Kashmir, adds Habib.[44]

---

[44] *Man and Environment: The Ecological History of India*, pp. 83-84.

In 1534[45], during the reign of Muhammad Shah, king of Kashmir, there occurred an acute famine due to the failure of crops in the Kashmir Valley. There was huge mortality on the one hand and exodus of people on the other. The state of things remained unchanged for ten months.[46] In 1576, during the reign of Ali Shah chak, Kashmir suffered from a devastating famine which was caused by a very severe snowfall when paddy-crops were still standing in the fields. Thousands died for want of food and many left the country. The King of Kashmir exhausted the resources of the state to alleviate the sufferings of his subjects, but the famine lingered on for three years in all horror and destructive.[47] In 1583-84, again as a result of a famine, prices of daily necessities of life in Kashmir shot up. On account of the dryness of the year all means of subsistence of many people were exhausted.[48] The next famine which was caused by the failure of rains began in 1595 and continued up to 1598. It affected a large portion of Northern India, especially Kashmir and Lahore.[49] In 1642,[50] Kashmir was put to great hardship as a result of severe famine caused by heavy and continuous rainfall that damaged the paddy crop leading to great scarcity of food. About 50,000 people left the country and migrated to Lahore.[51]

---

[45] Whereas, Hassan records that, it had taken place in the year of 1532. See, *Tarikh-i-Hassan*, Vol. 1st. p. 458; *Shams-ut-Tawarikh*, p. 444.

[46] Natural Calamities and the Great Mughals, p. 25.

[47] *Waqiat-i Kashmir*, p. 149; *Tarikh-i-Hassan*, Vol. 1st. pp. 458-459; *Shams-ut-Tawarikh*, pp. 444-445. But the main text used above is directly taken from *Natural Calamities and the Great Mughals*, p. 29.

[48] Natural Calamities and the Great Mughals, pp. 29-30.

[49] Natural Calamities and the Great Mughals, p. 30; Man and Environment: The Ecological History of India, p. 84.

[50] However, Azam Didamiri and Hassan ironically both record that: the famine has taken place in the year of 1647. See, *Waqiat-i Kashmir*, p. 229; *Tarikh-i-Hassan*, Vol. 1st. p. 460; *Shams-ut-Tawarikh*, p. 445.

[51] Inayat Khan, *Shah Jahan Nama*, (ed.), W. E. Begley and S Z. A. Desai, OUP, New Delhi, 1990, pp. 291-292. (Here after, *Shah Jahan Nama*); *Natural Calamities and the Great Mughals*, p. 38; *Man and Environment: The Ecological History of India*, p. 85. Though Agarwal points out 50,000 people migrated from Kashmir to Lahore, but Habib rightly reduces the number to 30,000 only.

Native historians, according to Lawrence, record nineteen great famines regarding which they give gruesome details, but the important fact on which they are all agreed is that the famines were caused by early snows or heavy rain occurring at the time when the autumn harvest was ripening. In 19th century, Lawrence informs, there have been two terrible famines, one known by the name of Sher Singh[52] which was caused by the early heavy autumn snow of 1831[53], and the other recent disaster, which again was similarly caused by continuous rains which fell from October 1877 till January 1878. It has been calculated that the population of Kashmir was reduced from 800,000 to 200,000[54] by the famine of Sher Singh and a flood which followed the famine destroyed many important irrigation works and permanently submerged large areas of valuable cultivation. In the famine of 1877-79 there was an enormous loss of life. Years have passed now since the last famine occurred, but the Kashmiri proverb, „Drag tsalih ta dag tsalih ne', which means that

---

[52] Sher Singh was subahdar (provincial governor) of Kashmir during the reign of Maharaja Ranjit Singh.

[53] Whereas, Hassan records that, in 1832, during the rule of Ranjit Singh and the subahdari of Sher Singh, Kashmir witnessed a terrible famine ever. Two major reasons have been pointed out by Hassan as the cause forthe occasioning of such a terrible famine in Kashmir. One, there was an early heavy snowfall—which devastated the agricultural production standing in the open fields yet to be harvested. Two, the men of Ranjit Sing wilfully confiscated whatever agricultural production was stored &available with the people—which increased the intensity of famine further. The situation rose to the level that, there started the practice of cannibalism. Which was something worst ever seen in Kashmir. See, *Tarikh-i-Hassan*, Vol. 1st. pp462-463; *Shams-ut-Tawarikh*, pp. 447-448. Here it bears to mention that, Hassan also informs us on having some more famines (not mentioned above) occasioned in Kashmir during the years: 1604, 1686-1689, 1723, 1731, 1746, 1755, 1765, 1813 and 1864 (see, table 4), the details of which have rarely been put on record by other modern historians. The one common major reason of occurring faminesduring the years identified by Hassan was the total failure of all crops due to the excessive untimely rain fall in Kashmir valley. For details, see, *Tarikh-i-Hassan*, Vol. 1st. pp. 457-465; *Shams-ut-Tawarikh*, pp. 443-449; *Waqiat-i Kashmir*, pp. 297, 436-38.

[54] Since Hassan doesn't mention the mortality number caused by the famine of Sher Singh. One wonders to notice that, whether the mortality number put on record by Lawrence is exact.

„the famine goes but its stains remain', is true in all senses and the country has not yet recovered from the awful visitation of 1877.[55]

Table 9.4: Famines/ Droughts of Kashmir between 1550 and 1900 A.D.

Famine Date	Summary
984 Hijri (1576 A.D.)	Kashmir witnessed a devastating famine due to the failure of crops during the government of Ali Shah Chak. The position was such that living inhabitants used flesh of dead bodies as part of their food.
992 Hijri (1584 A.D.)	As a result of famine prices of daily necessities rose up. All means of subsistence were exhausted by the state.
1004 Hijri (1595 A.D.)	Famine occasioned due to the failure of weather. Not any kind of damage reported.
1013 Hijri (1604 A.D.)	Famine occasioned due to the failure of weather during the reign of emperor Akbar. Many people died. Though Kashmir received grain from Punjab and Lahore but that remained insufficient.
1052 Hijri (1642 A.D.)	Due to the excessive untimely rainfall severe famine occasioned during the reign of Shahjahan. Thirty thousand (30, 000) people migrated to Lahore.
1098/1101 Hijri (1686/1689 A.D.)	During the reign of emperor Aurangzeb people suffered deeply due to the scarcity of crop production in Kashmir.
1136 Hijri (1723 A.D.)	During the reign of Muhammad Shah Gazi famine occasioned due to the excessive rain fall. It continued for six months. Many people died.
1144 Hijri (1731 A.D.)	During the reign of Muhammd Shah and the governorship of Aehteram Khan: crops in the field remained unripen which occasioned famine. Out of frustration people revolted against each other and the government.
1159 Hijri (1746 A.D.)	During the reign of Muhammad Shah and the governorship of Afrasiyab Khan: crops failed to grow and severe famine occasioned. Loot,

Contd...

---

[55] The Valley of Kashmir, p. 213. Also see, Tarikh-i-Hassan, Vol. 1st. pp. 467-468; Shams-ut-Tawarikh, p. 451.

	chaos and destruction became the scene of the valley. One-third population of the valley perished.
1169 Hijri (1755 A.D.)	During the sultanate of Ahmad Shah Durrani and the governorship of Sukh Jeevan Mil: due to the fall of excessive rains crops failed and famine occasioned.
1179 Hijri (1765 A.D.)	During the period of Hakim Nooruddin Khan: the scarcity of the crop production brought famine in the valley.
1229 Hijri (1813 A.D.)	During the time of Sardar Azeem Khan: the crops in the field didn't ripe and brought famine. Many people had to face nothing but death.
1248 Hijri (1832 A.D.)	An early heavy snowfall and the oppressing nature of the men-in-government occasioned a devastating famine in Kashmir during the rule of Ranjit Singh and the Subahdari of Sher Singh. It was the worst kind of famine ever occasioned in Kashmir. It reduced the population of the state from eight hundred thousand (800,000) to two hundred thousand (200,000).
1281 Hijri (1864 A.D.)	During the government of Ranbir Singh, the paddy plants in the fields remained unripen, due to which people of the state felt scarcity of rice later-on. In order to get away from the difficult situation, Maharaja Ranbir Singh spent hefty money in purchasing ready crops from Punjab and Lahore for the people of Kashmir.
1294/1297 Hijri (1877/1879 A.D.)	Again during Ranbir Singh's government, excessive rain fall caused floods which destroyed the ready-crop-plants in the fields yet to be harvested. Therefore due to the scarcity of food thousands of people had to face death till the government enabled itself to control the worsening situation of the inhabitants.

Sources: Shahjahan Nama, p. 38; Waqiat-i Kashmir, pp. 229, 297, 436-438; Tarikh-i-Hassan, Vol. 1st pp. 458-465; Shams-ut-Tawarikh, pp. 444-449;The Valley of Kashmir, p. 213.

9.2.5 Fires

Hassan furnishes detailed information on the spread of fires that did spread in Kashmir during the period under study. He says, in 1620, during the subahdari of Dilawar Khan, Skindarpora witnessed fire which almost burnt down eleven thousand (11,000) houses close to each other. The Jama Masjid nearby, completely burnt down again by second time. Thereafter, it was emperor Jahangir—who developed the Sikandarpora to life again.[56] Again, in 1675, during the rule of Iftkhar Khan, Kawdara witnessed fire which burnt down twelve thousand (12,000) houses, and emperor Aurangzeb constructed the Jama Masjid now by third time.[57] The tragedy of the spread of fires didn't end thereafter. According to Hassan, Kashmir witnessed the spread of fires during the years:1711, 1737, 1739, 1745, 1765-66, 1800-01, 1850-51, 1875 and 1878 (see, table 5). And, in these unfortunate years, Hassan remarks that, though the loss caused by these fires was irreparable, yet some of the concerned rulers re-developed the burnt villages and towns to life again.[58]

According to Hassan, the reasons for the spread of fires in Kashmir during the period under study were various. But, one major reason Hassan cites was the rivalry between ruling parties—of inside and those of intruded from outside. Their conflict and rivalry to dominate over each other often ended spreading up the fires in the Valley. For this reason, some of the calamitous events of fires occasioned would definitely be called were man-made calamities.[59]

---

[56] *Tarikh-i-Hassan*, Vol. 1st. p. 477; *Shams-ut-Tawarikh*, p. 458. However, according to Azam Didamiri, the fire did spread in the year of 1617. He also mentions one more calamitous event of fire which spread during the subahdari of Nawab Ahmad Baig Khan in the year of 1615. See, *Waqiat-i Kashmir*, pp. 199-200.

[57] Tarikh-i-Hassan, Vol. 1st. p. 477; Shams-ut-Tawarikh, p. 459; Waqiat-i Kashmir, pp. 277-78.

[58] *Tarikh-i-Hassan*, Vol. 1st. pp. 476-479; *Shams-ut-Tawarikh*, pp. 457-460.

[59] See,*Tarikh-i-Hassan*, Vol. 1st. pp. 476-479; *Shams-ut-Tawarikh*, pp. 457-460.

Table 9.5: Fires of Kashmir between 1550 and 1900 A.D.

Fire Date	Most Affected Area(s)	Summary
1029 Hijri (1619/20 A.D.)	Skindarpora	In Skindapora fire did spread during the government of Dilawar Khan. Almost eleven thousand (11000) houses burnt upto the Rajouri and Saraaf bridges.
1086 Hijri (1675 A.D.)	Kawdara	During the government of Iftkhar Khan: in Kawdara fire emerged and burnt almost twelve thousand (12,000) houses around.
1123 Hijri (1711 A.D.)	Saraaf Kadal/ Malchmar	During the vice-regency of Nawazish Khan: after the occasioning of floods, fire broke out which burnt upto twelve thousand (12,000) houses between Saraaf bridge and Malchmar.
1150 Hijri (1737 A.D.)	Srinagar	A rivalry between Abu Al Barkat and his Wazirs occasioned spreading of fire in the city which destroyed almost twenty thousand (20,000) houses around.
1152 Hijri (1739 A.D.)	Srinagar	During the government of Atiyat Ullah Khan, Nadir Shah deputed Fakhrudullah as Nazim of Kashmir. Fakhrudullah wasn't shown any obedience by other subahdars—which led to war between them. In anger and frustration, fifteen thousand (15,000) houses were set to fire at the centre of the city and its surroundings.
1158 Hijri (1745 A.D.)	Srinagar	Babar Khan became disobedient to the ruler Mansoor Khan and revolted against him. Bambas and Kistwaries jointly took the advantage of the situation. They plundered Kashmir and set fire to fifteen Mohallas around.
Not known	Zainakadal	Due to the occurrence of famine during the period Afarsaiyab Khan,

Contd...

		common people revolted against the grain merchants. The office bearers confronted the trouble shooters in Zainakadal—due to which twelve localities were set to fire around.
1199 Hijri (1765/66 A.D.)	Tankipora/ Habakadal	During the cruel-some period of Azad Khan, fire emerged in Tankipora and spread upto Habakadal. Around eight thousand (8000) houses destroyed.
1215 Hijri (1800/01 A.D.)	Saraaf Kadal Mohalla	During the period of Abdullah Khan, many villages came under the spread of fire and destroyed.
1267 Hijri (1850/51 A.D.)	Tankipora/ Zeendarsahib Mohalla	During the period of Gulab Singh, fire emerged in Tankipora, and upto Zeendarsahib Mohalla, it burnt around two thousand (2000) houses.
1292 Hijri (1875 A.D.)	Tankipora	During the government of Ranbir Singh, again fire emerged in Takipora and destroyed seven hundred (700) houses.
1295 Hijri (1878 A.D.)	Srinagar	During the period of Ranbir Singh: an occurrence of famine took place due to no-rainfall for six months in hot season. Almost every day houses were caught to fire in the city and the villages. In these days, fire emerged in Habakadal and upto Kazi-zada embankment destroyed one thousand (1000) houses in just two hours.

Sources: Waqiat-i Kashmir, pp. 199-200, 277-278; Tarikh-i-Hassan, Vol. 1st pp. 477-479; Shams-ut-Tawarikh, pp. 457-460.

9.3 CONCLUSION

The conclusion that we draw from the foregoing pages is that: calamitous events in Kashmir is not completely a new phenomenon, rather these had been happening there from ancient times too. However, the frequent happening of calamities (particularly, earthquakes, floods & fires) in recent times demands more concern. Because scientific research reveals that, these calamities happen due to the sudden shift in climate—which does occur due to the changes in the landscape. Kashmir, as the geo-scientists warn, is prone to calamities by nature. Thus any imbalance in its landscape ecology creates more chances of calamities to happen. For example, the available data for the period under study showcase that, calamitous events like floods and earthquakes have been terribly increasing in Kashmir. Therefore, from the research point of view, we can infer that, lot of physical changes have been happening in the landscape ecology of Kashmir—which directly or indirectly accumulate an imbalance in the ecosystem. This nuisance practice of ours continuing way back has a direct bearing on the calamitous events happen.

The large-scale human intervention in the landscape patterns more often than not continuing, say for instance, is: unregulated regular cutting down of mountain areas, testing & dumping of ammunition in the ecological zones, decomposition of abundance of over-dated articles in open fields of the forest areas, and continuous deforestation. All of which have thus brought negative prolonging impact on the ecology of Kashmir. For example, the precious biodiversity loss over the years includes the tremendous economic loss as well. To conclude, there is no denial of the fact that, unregulated interventions have been randomly speedily increasing in the ecological zone(s)which situates to be the root cause of frequently occasioning calamitous events in Kashmir. Thus to safeguard Kashmir from receiving frequent calamitous events like floods, fires and earthquakes: the governing body of environmental laws has to work on wiser conceptual models in order to be efficient functional.

Chapter 10

Environment Impact Assessment in India: Constitutional Perspective

Jaspal Singh & Varinder Singh

Abstract: *India is the first country, which has made provisions for the protection and improvement of environment in its Constitution by 42nd Amendment 1976. Environmental Impact Assessment is an important management tool for ensuring optimal use of natural resources for sustainable development. The process of environmental impact assessment (EIA) in India has been one of parallel evolution with the development of an institutional capacity to co-ordinate and monitors the environmental policies and procedures going hand-in-hand with the formulation and implementation of the necessary legislation and regulations. Various laws have been framed out in order to make out some clearances for environment protection. It is important to note that the demand for environmental assessment of industries was taken up as an executive order in 1994 through Ministry of Environment and Forests notification – without any legislation being passed, or the Parliament being involved. The prioritisation of development and the dilution of regulation have had predictable consequences. Although an assessment may lead to difficult economic decisions and political and social concerns, environmental impact assessments protect the environment by providing sound basis for effective and sustainable development. Today, NGO's and communities continue to struggle with fraudulent EIA reports, staged public hearings, and unscrupulous environmental clearances. As public opposition to specific projects has become increasingly intense – it is growing more evident that the questions that were passed by too quickly about environmental impact assessment in the beginning must be asked and answered again.*

Keywords: *Assessment, Constitution, Environment, Judiciary and Projects.*

10.1 INTRODUCTION

The conservation, protection and preservation of the environment is enshrined within the Indian Constitution which enjoins the "States to take measures to protect and improve the environment and to safeguard the forests and wildlife of the country". However, whilst the Indian Constitution was one of the first in the world to recognise the importance of environmental conservation, this awareness has not prevented India from encountering major environmental problems as it strives to achieve its development aims[1]. The technological state under the impact of the philosophy of welfarism does not eliminate, but intensifies the conflict between environmental values and development needs. Sustainable development is now the cornerstone of all policies and procedures relating to development activities in the country. Legal strategies are, however necessary to reconcile the conflict and to augment sustainable development. Environment Impact Assessment (EIA) is an instrument of reconciliation[2]. Environmental Impact Assessment can be broadly defined as the systematic identification, evaluation and monitoring of the potential impacts (effects) of proposed projects, plans, programmes or legislative actions relative to the physical – chemical, biological, cultural and socio-economic components of the total environment[3]. EIA is a legal planning tool that is now generally accepted as an integral component of sound decision making of whether to proceed with a project or not. The prime objective of EIA is to predict and address potential

---

[1] Will Banham and Douglas Brew, "A review of the development of environmental impact assessment in India" retrieved from <http://www.tandfonline.com/action/journalInformation?journalCode=tiap18> accessed on 15.12.2017.

[2] Jai Jai Ram Upadhyay. (2005). *Environmental Law.* Allahabad: Central Law Agency315.

[3] L.W. Canter. (1996). *Environmental Impact Assessment* (2nd ed.) McGraw-Hill Inc 241.

environmental threats at an early stage of project planning and design.

Environmental Impact Assessment can be characterized as the investigation to anticipate the impact of a proposed movement/venture on the environment. Environmental Impact Assessment is the assessment of the environmental consequences (positive and negative) of a plan, policy, program, or actual projects prior to the decision to move forward with the proposed action. The term "Environmental Impact Assessment" is usually used when applied to actual projects by individuals or companies and the term "Strategic Environmental Assessment" (SEA) applies to policies, plans and programs most often proposed by organs of the state[4]. Environmental assessments may be governed by rules of administrative procedure regarding public participation and documentation of decision making, and may be subject to judicial review. The purpose of the assessment is to ensure that decision makers consider the environmental impacts when deciding whether or not to proceed with a project. The International Association for Impact Assessment (IAIA) defines an environmental impact assessment as "the process of identifying, predicting, evaluating and mitigating the biophysical, social, and other relevant effects of development proposals prior to major decisions being taken and commitments made"[5]. EIAs are interesting in that they don't expect adherence to a foreordained environmental result, but instead they require leaders to represent environmental esteems in their choices and to legitimize those choices in light of Nitti gritty environmental investigations and open remarks on the potential environmental impacts. EIA isto evaluate and identify the predictable environmental consequences and the best combination of economic and environmental costs and benefits of the proposed project.[6]

---

[4] Retrievedfrom<http://www.nkc.ac.in/uploaded_files/study%20material_evs_unit_%20iv.pdf.>accessed on 10.12.2017.

[5] Satish C. Shastri. (2012). *Environmental Law*.Lucknow: Eastern Book Company125.

[6] Anji Reddy Mareddy.(2017). *Environmental Impact Assessment: Theory and Practice*. Hyderabad:BS Publication 5.

EIA systematically examines both beneficial and adverse consequences of the project and ensures that these effects are taken into account during project design. It helps to identify possible environmental effects of the proposed project, proposes measures to mitigate adverse effects and predicts whether there will be significant adverse environmental effects, even after the mitigation is implemented.

By considering the environmental effects of the project and their mitigation early in the project planning cycle, environmental assessment has many benefits, such as protection of environment, optimum utilization of resources and saving of time and cost of the project. Properly conducted EIA also lessens conflicts by promoting community participation, informing decision makers, and helping lay the base for environmentally sound projects. Benefits of integrating EIA have been observed in all stages of a project, from exploration and planning, through construction, operations, decommissioning, and beyond site closure. The purpose of EIA is not just to assess impacts and complete an environmental impact statement (EIS); it is to improve the quality of decisions. Through informing the public, the project proponent can make environmentally sensitive decision by being aware of a project's potential adverse impacts on the environment. Another purpose of EIA is to inform the public of the proposed project and its impacts.[7] In this context public participation provides crucial information. Through their participation, the project proponent will be able to take advantage of the information that citizens contribute concerning values, impacts, innovative solutions and alternatives.

10.2 LEGAL FRAMEWORK OF EIA IN INDIA

The Planning Commission of India in its Seventh Five Year Plan stressed the need for EIA in India as follows:

By the year 2000, industrialization of the country will have reached a stage when in the absence of effective remedial measures, severe problems of air, water and land Pollution will assume serious

---

[7] Bram F. Noble. (2015).Introduction to Environmental Impact Assessment: A Guide to Principles and Practice.London: Oxford University Press 21.

proportions.... In project planning, besides the availability of raw material, man power and funds, decisions regarding the use of the environment will have to be taken, investments built-in for minimizing environmental damage or degradation. This will apply equally to the public and private sectors. A new type of expertise in environmental impact analysis will have to be developed and applied for deciding the optimum location of any project.

After the United Nation's Conference on Human Environment held at Stockholm in 1972, nations of the world in general and signatories to this Conference in particular became serious[8] about taking necessary measures including legislative ones for the protection of the earth's natural resources; prevention, control and abatement of environmental pollution, and evolving principles of liability and compensation to redress the grievances of the victims of environmental pollution. India, being signatory to this Conference, did not lag behind and came up with certain Acts, Rules and Notifications to achieve these objectives. Not only this, but two very important Articles, *viz.,* 48-A and 51A(g) were added to the Constitution of India through the Constitution Forty Second Amendment Act, 1976, which make it obligatory for the State and citizens as well to protect the environment. From amongst these legislative measures, the Water (Prevention and Control of Pollution) Act, 1974, the Air(Prevention and Control of Pollution) Act, 1981, the Environment (Protection) Act,1986 along with the Rules made there under and the Environment Impact Assessment Notification of 1994 and 2006 are most relevant to the aspect of the protection of environment and prevention, control and abatement of environmental pollution.

10.3 LEGISLATIONS REGULATING ENVIRONMENTAL CLEARANCES

The ontology of environmental laws in India consists of resource conservation legislation, pollution control laws and dispute redressal regulatory mechanism. The EIA precepts and principles are

---

[8] Bhanu Pratap Singh (2013). EIA: Do we really need a shift. *International Journal of Environmental Engineering and Management*, Vol No. 4(Issue No. 3), 227-232.

recognized explicitly and implicitly within the corpus of environmental laws. One of the earliest legislation in this direction was the Indian Forest Act, 1927 which addresses social and ecological impact with an inherent bias of commercial exploitation of forest resources. The social impact assessment (SIA) provisions are generally discerned in the acquisition proceedings of forest land[9].

The Air (Prevention and Control of Pollution) Act, 1981 further amended in 1987 was legislated with the mandate to the Government for ensuring prevention and control of Air pollution including noise pollution. The Government was further mandated to set up and constitute pollution control boards at Central and State Government levels. No one could establish or operate any industry without the concerned Board's clearance.

The Water (Prevention and Control of Pollution) Act, 1974 and further amended in 1988 mandated the Government to ensure prevention of degradation of quality of water, restoration of water quality. No industry can be established or operated without clearance from the respective Pollution Control Boards. No developer is permitted to discharge sewage or industry effluents in any river/stream[10].

The Wildlife (Protection) Act,1972 and further amended in 1993 stipulates that any industry or other developmental activities (including road and highway projects) requiring prior EC in terms of the thresholds detailed in the EIA notification, Sept 2006shall have to obtain Wildlife Clearance if the project is proposed to be located inside of 10 kms. of any National Park/Wildlife Sanctuary.

The Forest Conservation Act, 1980 and further amended in 1988 aimed to empower the Central Government to ensure prevention of depletion of forest area. It prevents de-reservation of forest land and

---

[9] Nomani, Md. Zafar Mahfooz. (2010). *Environment Impact Assessment Laws*. New Delhi: Satyam Law International83.

[10] Ajit K. Sinha and K.N. Jha, "Environmental clearance Acts and Rules - Evolution and experience", retrieved from https://icjonline.com/views/2017.05.POV_KNJha.pdf> accessed on 10.11.2017.

to ensure that forest land is not used for non-forest purposes without the express approval of the Central Government.

10.4 PARADIGM SHIFT OF EIA

The primary aim of EIA procedures is to gauge the potential environmental impact of an economic project so as to allow for measures to minimize that impact. The methodology adopted is that of self-assessment by the project proponent followed by review and project approval by the regulators. The EIA notification was first issued in 1994 by the Central Government, Ministry of Environment and Forests (MOEF), in exercise of its power to take any measures to protect and improve the environment as provided under Section 3 of the Environment Protection Act, 1986. It introduced a process for prior environmental approval of certain kind of projects (specified in Schedule 1). The project proponents were required to submit an environmental assessment report, environmental management plan and the details of the public hearing conducted in the vicinity of the project (exceptions to these requirements were permitted for certain projects). The MOEF would function as Impact Assessment Agency which could consult a Committee of Experts set up for this purpose[11].

The EIA notification of 1994 was further improved upon with the more detailed notification of 2006. The latter provided for significant and re-engineered environment clearance process for effectively carrying out EIA. The provision comprised categorization of projects in terms of set out parameters. It categorises projects as A and B for the purpose of clearance by centre and state respectively. Such projects will be appraised at state level by constituting State Level Environment Assessment Authority (SEIAA) and State Level Expert

---

[11] Nupur Chowdhury , "Environmental Impact Assessment In India: Reviewing Two Decades of Jurisprudence"retrieved from
https://www.google.co.in/url?sa=t&rct=j&q=&esrc=s&source=web&cd=2&cad=rja&uact=8&ved=0ahUKEwjw7ILF_NTYAhWHRY8KHZ0VAtkQFggsMAE&url=http%3A%2F%2Fwww.iucnael.org%2Fen%2Fdocuments%2F1140-environmental-impact-assessment-in-india-reviewing-two-decades-of-jurisprudence%2Ffile&usg=AOvVaw2dNkDRAlRr8pDW_7nxzEzZ> accessed on 15.12.2017.

Appraisal Committee (SEAC) constituted by the Central Government in consultation with the State Governments/UTs Administration. In fact, the process was democratized and the 2006 provisions further ensured closer interaction between the State Governments and the Central Government.

Para 12 of the EIA Notification,2006 provides Interim Operational Guidelines Providing Inter-Link between EIA Notifications, 1994 and 2006 which states that Central Government may relax one or all provisions of EIA Notifications, 2006 except the activities, prior environment clearance in Schedule 1 of the Notification[12]. The Ministry has issued Interim Operational Guidelines (IOG) from time to time providing methodology for disposing of pending cases with Central and State Governments at various stages. Public Hearing proceedings are also better structured and time bound.

10.5 CONSTITUTIONAL FRAMEWORK OF EIA

With the spreading environmental consciousness, need has arisen to peep into the affairs of activity which affects or are likely to affect the environment. The Constitution of India has been providing crystal clear in its provisions a due consideration for the protection and promotion of environment. The current focus upon environmental issues is not something new but in fact something which has been part and parcel of our Indian culture. When the Constitution was drafted it did not contain any specific provision for the protection of environment and even the word „environment' did not find any place anywhere in the Constitution. The Constitution of India under the ambit of Fundamental Rights, Directive Principles of State Policy and Fundamental Duties puts the knot of environment on a very high pedestal. The 42nd Amendment 1974 came up with the responsibility of the State as well as the citizens to improve the environment and to safeguard the forest and wildlife of the country.

The process of environmental impact assessment (EIA) in India has been one of parallel evolution with the development of an institutional capacity to co-ordinate and monitors the environmental

---

[12] Nomani, Md. Zafar Mahfooz. (2010). *Environment Impact Assessment Laws.* New Delhi: Satyam Law International 275.

policies and procedures going hand in hand with the formulation and implementation of the necessary legislation and regulations. The decrease in the number of projects rejected between 1990 and 1994 might indicate a gradual improvement in the quality of EIA's[13]. This need was, however imperative in our democratic nation.

Under the Indian federal system, governmental power is shared between the Union and the State governments. Part XI of the Constitution governs the legislative and administrative relations between the Union and the States. The division of legislative powers shows that, there are ample provisions to make law dealing with environmental problems at the local level as well as at the national level, but under the federal system, the Central Government controls the finances largely[14]. It may happen that when an industrial project is allocated to a particular state, it may have some environmental impact in that State and thus it may be opposed by the environment and planning department of the state concerned. On the other hand, the Central Government may threaten to withdraw the project from the State if its implementation is opposed and resulting into a conflict between development and environment. This conflict is being taken care of by the Environment Impact Assessment (EIA) which is an effort to anticipate measures and weigh the socio-economic and eco- system changes that may result from the proposed project.[15] In *Adivasi Majdoor Kisan Ekta Sangathan v. Ministry of Environment and Forest*[16] the evidence of persons who voiced their opposition to the project was not recorded and no summary of the public hearing was prepared in the local language nor was it made public. Therefore, the Court declared the approval invalid.

---

[13] Will Banham and Douglas Brew, "Environmental Assessment" retrieved from <http://www.tandfonline.com> accessed on 12.01.2018.

[14] "The Indian Constitution and Environmental Protection" retrieved from <http://www.shodhganga.inflibnet.ac.in> accessed on 12.01.2018.

[15] *Ibid.*

[16] Appeal No. 3/2011 (T) (NEAA No. 26 of 2009). Judgment of Principal Bench of the National Green Tribunal on April 20, 2012.

Article 14 of the Constitution provides for a framework of „Right to Equality' and Article 19(1)(g) enshrines about the „Freedom to carry on any trade, occupation, business and profession'. These two provisions are to be read in consonance with each other. Under the ambit of Article 19(1)(g), every citizen is free to carry on any sort of trade, occupation, business and profession but however, this provision is clubbed with the reasonable restrictions as stated under Article 19(6). So, the freedom to start any business or any other activity should not be such as hampering the environmental protection. Any such activity, if results, in polluting the environment needs to get the clearance certificate from the authorities concerned about its environmental impact. So, it becomes imperative to assess the environmental impact arising out from such activity. The reason for this lies in Article 14, which provides for equality norm i.e. everyone has the right to equality to live in a healthy and unpolluted environment like others where such activities are not carried upon. The Indian Constitution guarantees „right to equality' to all persons without any discrimination. This indicates that any action of the State relating to environment must not infringe upon the right to equality as mentioned in Article 14 of the Constitution. The Stockholm Declaration, 1972 also recognised this principle of equality in environmental management and it called up all the worlds' nations to abide by this principle[17]. So, the constitutional provisions provide us with an avert call to abide by the notion of environment impact assessment.

Article 21 of the Constitution of India provides for „Right to Life and Personal Liberty' which in itself imbibes the core aspect of environment protection as well. The provision provides for a right to live in a healthy environment[18]. Environment impact assessment provides for evaluating the expectancy of environmental impacts arising or likely to arise from some proposed project or development. If it becomes the duty on one side to evaluate the assessment, it in turn maintains a balance between providing the persons with their right to live in a healthy and clean environment. According to

---

[17] "The Indian Constitution and Environmental Protection" retrieved from <http://www.shodhganga.inflibnet.ac.in> accessed on 12.01.2018.

[18] Subhash Kumar v. State of Bihar, AIR 1991 SCC 598.

Bhagwati J., Article 21 "embodies a constitutional value of supreme importance in a democratic society". Krishna Iyer J. has characterised Article 21 as "the procedural magna carta protective of life and liberty". The ‚right to life' under Article 21 means a life of dignity to live in a proper environment free from the dangers of diseases and infection. Maintenance of health, preservation of the sanitation and environment have been held to fall within the purview of Article 21 as it adversely affects the life of the citizens and it amounts to slow poisoning and reducing the life of citizens because of the hazards created if not checked[19]. The same can be the result of the activities or projects undertaken which are affecting or likely to affect the environment if not properly assessed with regard to their environmental impact. So, in order to protect and preserve the right to life of any person, it has become the need of an hour to assess the impact which any activity is likely to have on the environment. In the landmark judgment[20] of Supreme Court, the Court took cognizance of the environmental problems being caused by tanneries that were polluting the water resources, rivers, canals, underground water and agricultural land. Therefore, the court issued several directions to deal with this problem. In *Rural Litigation and Environment Kendra, Dehradun v. State of Uttar Pradesh*[21], the representatives of the Rural Litigation and Entitlement Kendra, Dehradun wrote to the Supreme Court alleging that illegal lime stone mining in the Mussorie-Dehradun region was causing damage to the fragile eco-systems in the area. The Court treated this letter as a public interest petition under Article 32 of the Constitution. And also, several committees have been appointed for the full inspection of illegal mining sites. All the committees came at the conclusion that the lime stone quarries whose adverse effects are very less, only those should be allowed to operate but that too after further inspection. Therefore, the court ordered the closure of a number of limestone quarries. Although, the Court did not mention any

---

[19] Drishti, "Article 21 of the Constitution of India – Right to Life and Personal Liberty" retrieved from <http://www.lawctopus.com/academike/> accessed on 12.01.2018.

[20] Vellore Citizens Welfare Forum v. Union of India, AIR 1996 SC 2721.

[21] [1998] INSC 254.

violation of fundamental right explicitly but as impliedly admitted the adverse effects to the life of people and involved a violation of Article 21 of the Constitution. It is crystal clear from this judgment of our apex court that the court gave guidelines with a view to assess the impact of any activity (mining of lime stone) over the environment.

The Constitution of India provides for the responsibility at the State level to make some policies which thereby gives directions to the State to make laws keeping in mind few of the indispensable principles. Some of the Directive Principles of State Policy showed a slight inclination towards environmental protection like: Article 39(b), 47, 48 and 49 individually and collectively impose a duty on the State to create conditions to improve the general health level in the country and to protect and improve the natural environment. Article 47 of the Constitution of India is considered to be more important as it imposes primary duty on the State to provide public with improved health, raised level of nutrition and ultimately improved standard of living. However, public health can be assured to the public only by offering the safe and protected environment to live in. This enabled the framers of the Constitution to be more conscious on the environmental concern. Article 48-A of the Constitution of India states:

The State shall endeavor to protect and improve the environment and to safeguard the forests and wildlife of the country.

This provision tends to place burden on the State to make such laws and provisions that can provide for the welfare of the environment. Such laws are to be made which maintain a balance between the state's activity to plan industrial projects and the right to everyone to live in healthy and unpolluted environment. Therefore, the balance can only be maintained if the results of such activities are assessed over its impact on environment. This assessment of activities for its impact over environment can however, be made a state responsibility to peep into this matter. The Environment Impact Assessment should be, therefore, prepared on the basis of the existing background pollution levels *vis-a-vis* contributions of pollutants from the proposed plant.

On the jurisprudential notion, we know that rights and duties co-exist. Therefore, the same goes with the Constitution. Fundamental Rights under the Constitution runs hand in hand with Fundamental Duties enshrined therein for the promotion and protection of the environment. Article 51-A (g) of the Constitution of India provides for the fundamental duty with regard to environmental protection. In the case of *Vijay Bansal v. State of Haryana*[22], the court held that before auctioning the mining areas falling within the fragile Shivalik ranges of Himalayas located in more than one District, the State of Haryana shall be under a legal obligation to apply to the Expert Appraisal Committee for the preparation of Environmental Impact Assessment Report. In the latest landmark judgment of *Sandeep Shah v. State of Rajasthan*[23], the court held that the State of Rajasthan is restrained from granting any further mining leases, quarry licenses and short-term permits including the renewal of existing mining leases/quarry license/short term permits, except only after getting the Environmental Impact Assessment Authority Report. The fundamental duty under Article 51-A (g) is the duty of citizen as to not to work upon any such activity which tends to deteriorate the environment.

In a democratic and developing country like India, growth and advancement is much needed. But the development of the nation should not be done at the cost of environment. The notion of Sustainable Development comes there from because no development can be done at the cost of destructing the environment. Thence, it can be enunciated that Environmental Impact Assessment is in consonance with the issue of Sustainable Development. In a case[24] decided by National Green Tribunal, Environment Protection Act, 1986, Sections 3(2) and 2(3)(v) was involved. The operation of Calcined Petroleum Coke Plant was initiated without obtaining environmental clearance. Entry 4(b) of Environment Assessment Notification 2006 requires mandatory prior environmental clearance

---

[22] AIR 2009 P&H HC.

[23] AIR 2015 Raj. HC

[24] Mamata Samantaray Chairperson State Progressive Women's Forum v. Union of India, NGT, Eastern Zone, Kolkata Bench, 23.12.2015.

before operation/construction/expansion/modernization when its capacity is 25000 MT or more. Therefore, the Tribunal ordered that of operational capacity increased above 25000 MT, steps for closure of the unit shall be undertaken. In a landmark authority of *Sterlite Industries (India) Ltd. v. Union of India*[25], the Supreme Court discussed the specific grounds on which administrative action involving the grant of environmental approval could be challenged. The grounds for judicial review were illegality, irrationality and procedural impropriety. Thus, the granting of environmental approval by the competent authority outside the powers given to the authority by law would be grounds for illegality. If the decisions were to suffer from Wednesbury unreasonableness, the Court could interfere on grounds of irrationality. Last, an approval can be challenged on the grounds that it has been granted in breach of proper procedure. Nevertheless, the Court has not restrained itself, in cases where it found that the SEAC had recommended approvals without any application ofmind. In another case of *M.C. Mehta v. Union of India*[26], cutting of a large number of trees is involved and it will be appropriate if the question of grant of the clearance having regard to the environmental impact is examined by the Forest Advisory Committee constituted by the Ministry of Environment and Forests. Hence, based on the report of the Environment Impact Assessment Authority of U.P. to the Chief Engineer of the U.P. Public Works Department, clearance certificate was issued. In another landmark judgment[27] of the Supreme Court, public interest litigation was filed in favour of environmental protection. The court concluded its decision by cancelling of Jal Mahal Tourism Project which was made by Mansagar Lake Precincts Lease Agreement. In *Centre for Social Justice V Union of India & Others*[28], Justice M.S. Shah of the Gujarat High Court dealt in detail with the Environment Impact Assessment Notification of 1994 as amended in 1997 and gave certain directions to the state of Gujarat about the manner in

---

[25] 2013 AIR SCW 3231.

[26] AIR 2008 SC.

[27] Jal Mahal Resorts P. Ltd. v. K.P. Sharma, AIR 2014 SC 134.

[28] AIR 2001 Guj. 71.

which a meaningful public hearing should be conducted before granting environmental clearance certificate to any industry, operation or process. In another case of *Gram Panchayat Navlakh Umbre v. Union of India and Ors*[29], the Court held that the "decision making process of those authorities besides being transparent must result in a reasoned conclusion which is reflective of a due application of mind to the diverse concerns arising from a project such as the present. The mere fact that a body is comprised of experts is not sufficient a safeguard to ensure that the conclusion of its deliberations is just and proper." In *V. Srinivasan v. UOI*[30], the role of private expert bodies and consultants conducting the EIA has also attracted judicial scrutiny. Furnishing of false information by the consultant has been deemed by the Court professional misconduct and it has recommended strict action in such cases.

It can be articulated that Indian judiciary has also taken initiative and commendable steps to accentuate the protection of environment. The State's responsibility does not wean off after making the provisions for protection of environment but however, the actual role begins when the provisions are to executable. In this regard, the provisions are to be implemented in the light of Environmental Impact Assessment, for the reason, that the environment can be protected and provisions can be better administered only if it fulfils that criteria of safeguarding the environment with the policies made by the Government. In India, the need for Environmental Impact Assessment has been recognised even by the Planning Commission by the Seventh Five Year Plan. However, existing system of administrative framework with its centralized environmental appraisal may lead to conflict between the project authorities and environmental authorities.

---

[29] Public Interest Litigation No. 115 of 2010. Judgment of Bombay High Court on June 28, 2012.

[30] Appeal No. 18 of 2011 (T), Judgment of Principal Bench of the National Green Tribunal on February 24, 2012.

10.6 CONCLUSION

While development is inevitable and essential to improve the quality of life, to meet basic human needs and secure better prospects for the citizens of developing countries, it is also equally essential to ensure that development takes place on a sustainable basis. The Government of India vide its policy and regulatory functions, ensures protection and improvement of the environment of India. The Constitution of India vide the 42nd amendment, has detailed and outlined this onus and responsibility on the Government of India in Articles 48-A and 51-A (g) of the Indian Constitution. The biggest problem facing India's environment is not a lack of environmental laws. Nor is it a lack of precedent to protect our environment. The single biggest issue facing India's beleaguered, yet resilient environment today is the failure of the Indian Government to adequately enforce existing environmental laws. There is no excuse good enough, no obstacle obtrusive enough, and no circumstances restrictive enough to exonerate the government from failing to perform its statutory duty to arrest environmental declines.

Presently, EIA is the only environmental tool which legally ensures that any new project is launched/installed/setup in such a way that it causes least damage to the environment safeguards the right to healthy environment. Environmental assessment enables us in carrying out environmental cost-benefit analysis of projects at an initial stage. It is thus, a precursor to detailed analysis of environment impacts, which are taken up only if a need for the same is established. It gives a view of the actors involved in the ,development-environment linkages'. However, the assessment process of impacts on the environment and affected people is very complicated, vast and varied as well as interdisciplinary in nature. Further, it is very difficult for, even experts in the field, who attempt at estimating the costs and benefits of these impacts. The difficulty starts with establishing the baseline values of the existing environment as well as estimating the change in these values on account of the project/activity.

The need for greater transparency, ensuring accountability of regulators and improving the quality of public participation are some of the challenges related with the present norms of EIA. It isn't lack of knowledge, therefore, that is responsible for the persistence of these problems. Correcting these systemic issues requires, first, the recognition of two critical dimensions of developmental decisions: (i) the social impacts of the breakdown of the primal relationship between a land and its people are multifarious and far-reaching, and require more serious consideration than has been given hitherto in the „cost-benefit approach' to impact assessment; and (ii) the power of incentives, both intended and perverse, for different categories of stakeholders.

An effective public participation programme does not happen by accident, it must be carefully planned. Public input can be a crucial and valuable source of expertise before, during and after the project planning and decision making. Thus, in order to realize the fundamental right to healthy environment, the Environment Impact Assessment, in strict sense of the term, must be more explicit in defining the affected area according to potential socio-economic impacts. If there is political will to recognise these, there is every likelihood of positive action towards building a more meaningful public consultation and review process - and thereby, more meaningful, and sustainable development.

Chapter 11

Environment Impact Assessment Principles Under Land and Heritage Conservation Laws: An Enviro-Legal Analysis

Md. Zafar Mahfooz Nomani

Abstract: Environment Impact Assessment (EIA) being a tool of sustainable development necessitates a vibrant legal policy for incorporation of man-made, natural and cultural heritage conservation. The EIA law in India is an off shoot of Environment Protection Act, 1986 and exists in the form of delegated legislation. The heritage conservation has escaped the attention of the legislature in framing the EIA law. The paper makes a case for refurbishing the legislative regime of EIA and heritage conservation by undertaking an in depth analysis of international and national heritage conservation laws to assess the course and direction of EIA law. The classical roots of the heritage conservation Laws despite its narrow focus can become a handy guide to promote rudimentary EIA principles. Concurrently the EIA law should also move beyond the desideratum of heritage conservation and come closer to biodiversity, cultural diversity and sustainable development.

Keywords: Heritage conservation, cultural heritage, natural and man-made heritage, environment impact assessment, marine and coastal heritage, continental shelf, maritime zone, natural and cultural biodiversity

11.1 EIA AND HERITAGE CONSERVATION

The heritage conservation laws are seldom applied for the environmental safeguards and scaling of environmental assessment in India. The seminal importance of land and heritage conservation legislations in promoting of short term, long term and disaster prone situations are rarely assessed logically to apply them for national and transnational environmental impacts. The concerns for the protection of cultural, man-made and natural heritage were raised in international law right from the 20th century. The rudimentary approach reflected in as far back as in 1907 in the Hague Regulations, 1907 protect the historic monument from the war like situations such as sieges and bombardments.[1] The United Nations Educational and Scientific Organizations (UNESCO) as a specialized agency has been seriously addressing the heritage conservations since 1950's.The UNESCO,,s Convention on Protection of Cultural Property in the event of Armed Conflict ,1954 is among the earliest known conventions of the modern international heritage conservation. The Convention popularly known as Hague Convention ordained the international law governing the protection of cultural heritage began comparatively narrow objective of with in a war time measures.[2] The Recommendation on International Principles Applicable to Archeological Excavation, 1956[3] adopted to streamline the modalities for the archeological excavations. The Recommendation, despite its limited subject-matter, the Recommendation contained principles fundamental to the protection of cultural heritage and international co-operation. The Recommendations was followed by Accessibility of Museums Convention, 1960[4] , the Landscape and Site Convention, 1962[5] and

---

[1] Hague Regulations Concerning The Law and Custom of War on Land to Protect Historic Monument From Sieges And Bombardments, 1907.

[2] Preamble to UNESCO,,s Convention on Protection of Cultural Property in the event of Armed Conflict, 1954

[3] The Recommendation on International Principles Applicable to Archeological Excavation, 1956 [UNESCO 5dec.1956]

[4] Recommendation Concerning The Most Effective Means of Rendering Museum Accessible to Everyone, 1960[UNESCO 14 Dec.1960]

[5] Recommendation Concerning The Safeguarding of The Beauty and Character of Landscape and Sites,1962 [UNESCO 11 Dec.1962]

Preservation of Endangered Cultural Property Convention, 1968.[6] Later on two significant conventions were adopted by UNESCO in the early 1970s, namely Convention on Trafficking on Cultural Property, 1970[7] Convention on Cultural Property, 1970[8] and the Convention on Protection of World Cultural and Natural Heritage (1972).[9] The growing body of the international instrument on cultural heritage is undergoing the constant epistemological refinement and interpretation.[10] This is evidenced from promulgations of the of the five Recommendations adopted between 1972 and 1980 by UNESCO.

The present paper is an attempt to locate environment impact assessment principle in land and heritage conservation laws to inform enlightened discourse and academic engagement in Indian context. This area seems to be more broadened in the sphere of EIA Law and land law to incorporate thematic thrust of heritage conservation within the purview of international and national environmental law.

11.2 EIA UNDER LAND LAWS

The land law of any country happens to manifests the legal and civilisational growth of a nation. The primary concerns of EIA principles need underlining in the basic land laws of India to draw a framework for the environmental and heritage conservations concerns. The Indian law on the subject is still reeling under the colonial legacy and doctrinaire limits of imperialist approach. Though the in the independent India modifications and amendments are brought in to reflect the nationalist conscience but

---

[6] Recommendation Concerning The Preservation of Cultural Property Endangered by Public Work or Private Work, 1968 [UNESCO 19 Nov.1968].

[7] Convention Relating to The Prohibition and Prevention of Trafficking in Cultural Property, 1970

[8] Convention On The Means Of Prohibiting And Preventing The Illicit Import ,Export And Transfer Of Ownership Of Cultural Property,1970 [14 Nov 1970[823U.N.T.S.231]

[9] Convention Concerning the Protection of the World Cultural and Natural Heritage,1972[16 Nov.1972,11 I.L.M.1358]]

[10] Janet Black, *International & Comparative Law Quarterly,* 'On Defining The Cultural Heritage', 49(1),61(2000)

the fact of the matter is that it has yet to reach to prescriptions of the environmental and heritage conservation laws. The Land Acquisition Act, 1894[11] is concerned with EIA to the extent of acquisition for public interest,[12] industrial development[13] infrastructural adequacies[14] and consequential impact on the environment. Invariably all the developmental project begins with the process of acquisition of land,[15] it seems desirable to explore the principles of EIA SIA and social impact under the law. Under the Act land may be acquired for public purpose as for companies.[16] The purposes for which acquisition of land for companies may be made are, however, restricted.[17]

> "Such acquisition may be made obtaining land (a) for the construction of dwelling-houses for workmen for the provision of amenities directly connected therewith; or (b) for the construction of some work which is likely proved useful to the public".[18]

---

[11] The *Land Acquisition Act,* 1884 [Act of 1894]: The text of the Act reflects the law as on 31.3.1996.
[12] *Id.* Section 6.
[13] *Id.* Section 44B.
[14] *Id.* Section 44.
[15] *Id.* Section 4 & 5A.
[16] *Id.* Section 38A: Acquisition of Land For Companies Industrial *Concern* To Be *Deemed* Company *For Certain Purposes:* An industrial concern, ordinarily employing not less than one hundred workmen owned by an individual or by an association of individual and not being a company, desiring to acquire land for the certain dwelling house for workmen employed by the concerned or for the provision of amenities directly connected therewith shall, so far as concerns the acquisition of such land, be deemed to be a company for the purposes of this part, and the references to company in [Sections 4, 5A, 6, 7 and 50] shall be interpreted as references also to such concern. [a] inserted by the *Land Acquisition (Amendment) Act* 1933 (XVI of 1933) see 6. [b] Substituted for see 5Am 6,7,17 and 50 by *Land Acquisition (Amendment) Act* (68 of 1984) see 22].
[17] *Id.* Section 40
[18] *Id.* Section 41

The legality of the Act came up for consideration before the Supreme Court in R.L. Aurora V. State of Uttar Pradesh.[19] The court held the fact that land can be acquired under Section 40(1)(b) read with Section 41 if the work constructed is for public interest and the public will be able to benefit from it in accordance with the terms and agreement. The act of land acquisition is important in various developing countries to foster community development and sustenance.

Doubts have arisen as to the validity of such acquisition. Some State governments have represented that the decision of the Supreme Court may have far reaching consequences in respect of acquisition of land for companies. It is feared that decision may render planned development of industries extremely difficult and also that there will be danger that the acquisition of land made for companies in the past might be questioned in courts of law and claims may be made by previous owners whose land have been acquired for restoration of land or payment of damages.

> "Unfortunately in a relentless pursuit of industrialization the Land Acquisition (Amendment) Ordinance, 1962 (3 of 1962) was promulgated on the 20th July 1962, suitably amending Sections 40 and 41 of the Act and also validating all past acquisition of land made for companies without any inbuilt mechanism for social and environment considerations".[20]

In acquisition of land for big projects, the practice generally followed under the Act, is that single notification is issued[21] under Section 4(1) of the Act which indicates that a particular area of land is needed or is likely to be needed for a public purpose. This is then followed by one or more declarations under Section 6 of the Act in respect of the land specified in the aforesaid notification to the effect that such land is needed for a public purpose or for a

---

[19] AIR 1962 SC 764
[20] The *Land Acquisition Amendment* Act, 1962
[21] *Id.* Section 4

company, as and when the plans are completed for the various stages of the project, e.g. plant, township and ancillary requirements.[22]

> To overcome the adverse effect of the Supreme Court judgment and in view of the urgency of the situation affecting many important projects, Land Acquisition Act, 1894, (...)Ordinance, 1967, on the 20th January 1967 to provide for submission of either one report in respect of the land which has been notified under Section 4(1) or different reports in respect of different parcels of such land to the appropriate (...)4(1) or of any land in the locality, as the case may be. The Ordinance specifically provides that, if necessary, more that one declaration may be issued from time to time (...)under Section 4, sub-section (1) of the Act irrespective of the fact whether one report or different reports has or have been made under section 5A, sub-section (2) of the Act.[23]

Thus the enormous expansion of the States' role in promoting public welfare and economic development since independence acquisition of land for public purposes, industrialization, building of institutions etc., has far more numerous than ever before. While this is inevitable, promotion of public purpose has to be balanced with the rights of the individual and his means of livelihood, again, acquisition of land for private enterprises ought not to be placed on the same footing as acquisition for the State for an enterprise under it.[24]

---

[22] The *Land Acquisition Amendment Act*, 1967 [The Act 13 of 1967]

[23] *Id.* Para 4

[24] Section 39: Previous consent of appropriate government and execution of agreement necessary: The provisions of [Sections 6 to 16 (both inclusive) and Sections 18 to 37 (both inclusive] shall not be put in force in order to acquire land for any company [under this part] unless with the previous consent at the

The individual and institutions who are unavoidably to be deprived of their property rights in land need to be adequately compensated for the loss keeping in view the sacrifice they have to make for the larger interests of the community.[25]

It is necessary, therefore, to restructure the legislative framework for acquisition of land so that it is more adequately informed by this objective of serving the interests of the community in harmony with the rights of the individual. Keeping these objects in view and considering the recommendations of the Law Commission, the Land Acquisition Review Committee as well as the State Governments, institutions and individuals, proposals for amendment to the Land Acquisition Act, 1894, were formulated and a Bill for this purpose was introduced in the Lok Sabha on the 30th April, 1982.[26]

The main proposals for amendment are as follows:

a) The definition of „public purpose' as contained in the Act is proposed to be amended as to include a longer illustrative list retaining, at the same time, the inclusive character of the definition.

b) Acquisition of land for non-Government companies under the Act will henceforth be made in pursuance of Part VII of the Act in all cases.

c) A time-limit of one year is proposed to be provided for completion of all formalities between the issue of the preliminary notification under Section 4(1) of the Act and the declaration for acquisition of specified land under Section 6(1) of the Act.[27]

It is believed that under these provisions a mechanism for social and environmental impact assessment can be suitably adopted by the collector. The law does express the concerns for public purposes, community welfare and balancing of corporate interests with the people. At this juncture the acquisition of land and

---

[appropriate government], nor unless the company shall have executed the agreement hereinafter mentioned.

[25] Id. [The Act 68 of 1967].

[26] Id. Para 2

[27] Ibid

preliminary inquiry and detailed EIA is made mandatory to auger sustainable development under the framework of the basic land laws of the country. This has not been adequately addressed in the past and the Act carries a very controversial image in not only marginalizing the traditional rights of community but adverse environmental impacts in the country.

11.3 CULTURAL HERITAGE PROTECTION LAWS

Most of the northern states of the world have fashioned their EIA laws around heritage conservation legislations. It seems appropriate to see the potentials of Indian heritage law to promote EIA principle. Under Indian context, the corpus of classical heritage conservation laws can be studied by analyzing the two major enactments on the subject viz, Ancient Monument Preservation Act, 1904 and Ancient Monuments & Archaeological Sites & Remains Act, 1958 to carve out the spaces for the cultural heritage protection laws in India.

11.3.1 ANCIENT MONUMENTS PRESERVATION ACT, 1904:

The heritage conservation law has become trend setting legislation in popularization of EIA principle. Indian heritage law obliquely addresses EIA under the Ancient Monument Preservation Act, 1904. The Act basically provide for the preservation of ancient monuments and for the protection and acquisition in certain cases of ancient, archaeological, historical and artistic interest.[28] To give effect to the EIA principles the Act empowers the Central Government[29] to control mining operation having an adverse impact on the monument. If the Central government is of opinion that mining, quarrying, excavating, blasting and other operations of a like nature should be restricted, or regulated for the purpose of protecting or preserving any ancient monument, then the

---

[28] The *Ancient Monument Preservation Act*, 1904 [The Act VII of 1904]: The text of the Act reflects the law as on 15.4.1969.

[29] *Id.* Preamble

Government may do so, by notification in the Official Gazette rules for the following:
  a. Fixing the boundaries of the area to which the rules are to apply;
  b. Forbidding the mining, quarrying excavating, blasting or any operation of a like nature except in accordance with the rules and with the terms of a license, and
  c. Prescribing the authority by which and the terms on which, licenses may be granted to carry on any of the said operations.[30]

The power to make rules given by this section is subject to the condition of the rules being made after previous publication. A rule made under this section may provide that any person committing a breach thereof shall be punishable with fine which may extend to two hundred rupees. If any owner or occupier of the land included in a notification under sub-Section (1) proves to the satisfaction of the (Central Government) that he has sustained loss by reason of such land being so included the (Central Government) shall pay compensation in respect of such loss.[31] Under the Amending Act of 1932 the Act Specifically makes provision for EIA by protecting of place of worship from misuse, pollution or desecration: It enjoins the authorities that if can by take such other action as he may think necessary in this connection.[32]

11.3.2 ANCIENTS MONUMENTS AND ARCHAEOLOGICALSITES AND REMAINS ACT, 1958

The Ancient Monuments & Archaeological Sites & Remains Act, 1958 incorporates the EIA principle in the rudimentary stage.[33] It defines the heritage in the context of ancient monument to mean any structure, erection or monument, or any tumulus or place of interment, or any cave, rock sculpture, inscription or monolith, which is of historical, archaeological or artistic interest and which

---

[30] The Ancient Monument Preservation (Amendment) Act, 1932.
[31] *Id.* Section 10A.
[32] *Id.* Section 13.
[33] The *Ancient Monuments and Archaeological Sites and Remains Act,* 1958 [The Act of the XXIV of 1958]: The text reflects the law as on 15.4.1969.

has been in existence for not less than one hundred years, and includes:

    a. The remains of an ancient monument,
    b. The site of an ancient monument,
    c. Such portion of land adjoining the site of an ancient monument as may be required for fencing or covering in or otherwise preserving such monument, and
    d. The means of access to, and convenient inspection of an ancient monument.[34]
    e. Whereas archaeological site and remains means any area which contains or is reasonably believed to contain ruins or relics of historical or archaeological importance which have been in existence for not less than one hundred years, and includes-
    f. Such portion of land adjoining the area as may be required for fencing or covering in or otherwise preserving it, and
    g. The means of access to, and convenient inspection of the area.[35]

The EIA is partially discernable it protective clause which reads that a protected monument maintained by the Central Government under this Act which is a place of worship or shrine shall not be used for any purpose inconsistent with its character. Where the Central Government has acquired a protected monument under Section 13, or where the Director-General has purchased, or taken a lease or accepted a gift or bequest or assumed guardianship of a protected monument under Section 5 and such monument or any part thereof is used for religious worship or observance by any community, the Collector shall make due provision for the protection of such monument or part thereof, from pollution or discretion-

    a. by prohibiting the entry therein, except in accordance with the conditions prescribed with the concurrence of the persons, if any, in religious charge of the said monument or part thereof, of any person not entitled so to enter by the religious usages of the community by which the monument or part thereof is used, or

---

[34] *Id.* Section 2(a).
[35] *Id.* Section 2(d).

b. By taking such other action as he may think necessary in this behalf.[36]

In order to protect the heritage sites from adverse environment impact the Act restrict on enjoyment of property rights to protected areas. It says no person, including the owner or occupier of a protected area, shall construct any building within the protected area or carry on any mining quarrying, excavating, blasting or any operation of a like nature in such area, or utilize such area or any part thereof in any other manner without the permission of the Central Government. Provided that nothing in this sub-section shall be deemed to prohibit the use of any such area or part thereof for purposes of cultivation if such cultivation does not involve the digging of not more than one foot of soil form the surface. The central Government may, by order, direct that any building constructed by any person within a protected area in contravention of provisions of sub-section (1) shall be removed within a specified period and, if the person refuses or fails to comply with the order, the Collector may cause the building to be removed and the person shall be liable to pay the cost of such removal.[37] The Act in the penal provision envisages that whoever- destroys, removes, injures, alters, deters, defaces, imperils or misuses a protected monument, or and does any act in contravention of sub-section (1) of Section 19, shall be punishable with imprisonment which may extend to three months, or with fine which may extend to five thousand rupees, or with both.[38]

11.4 MARINE AND COASTAL HERITAGE LAWS

The coastal heritage and marine biodiversity conservation is obliquely referred under marine pollution, territorial waters, continental shelf, exclusive economic zone (EEZ) *and other maritime zones laws under two major enactments on the subject viz; the Merchant Shipping Act, 1958 and Maritime*

---

[36] *Id.* Section 16.
[37] *Id.* Section 19.
[38] *Id.* Section 30.

Zones Act, 1976. These laws owe its origin to India's accession to the, *International Convention on Civil Liability for Oil Pollution Damage, 1969 and the Protocol to the Convention,* 1976.[39] A detailed analysis of these legislations is undertaken in the lines to be followed.

11.4.1 MERCHANT SHIPPING ACT, 1958

The Merchant Shipping Act, 1958 is one of important legislations to incorporate the national and transnational environmental impact specially the EIA of marine and aquatic resources of the country. The Act was passed to give effect to India's accession to the, International Convention on Civil Liability for Oil Pollution Damage, 1969 and the Protocol to the Convention, 1976. Under the Convention of 1969, the liability of the owner of the ship in respect of any one incident is limited to an aggregate sum of 2,000 Farce for each ton of the ships tonnage subject to a maximum of 210 million francs.[40] The 1976 Protocol, inter alia, replaced the reference to franc by special drawing rights. This was done to avoid problems created by the fluctuations in currencies of different countries in relation to frame and in conformity with the practice followed in most of the similar international Conventions.[41] This assumes significance as the pecuniary liability has a salutary impact on the EIA of shipment and navigation industries. Under the Amending Act, 1970 the civil liability for oil pollution damage is made applicable to every Indian ship wherever it is; and every foreign ship while it is at a port or place in India or within the territorial waters of India or any marine areas adjacent there to over which India has, or may hereafter have, exclusive Jurisdiction in regard to control of marine pollution under the territorial waters, continental shelf, exclusive economic zone (EEZ) and other Maritime Zones Act, 1976, or any other law

---

[39] *See:* The Statement of Objects And Reason, *Merchant Shipping Act,* 1958 Para 1.
[40] *Id.* Para 2.
[41] *Id.* Section 352 G.

or time being in force[42] shall be liable for any pollution damage caused by oil escaped from the ship as a result of the incident.[43]

Liability under Section 352-I in respect of any incident is limited to an aggregate amount of one hundred and thirty-three special drawing rights for each ton of the ship's tonnage; or fourteen million special drawing rights[44] whichever is lower. Any owner desiring to avail the benefit of limitation of his liability under sub-section (1) of Section 352J shall make an application to the High Court for constitution of a limitation fund. Such fund may be constituted either by depositing with the sum the High Court or by furnishing bank guarantee or such other security as, in the opinion of the High Court, is satisfactory. The amount is Special Drawing Rights to be deposited or secured in the fund under sub-section (1) shall be converted in rupees on the basis of official value in rupees of the special drawing rights as determined by the Reserve Bank of India on the date of constitution of the fund.[45] The owner of every Indian ship which carries 2000 tons or more oil in bulk as cargo shall in respect of such ship, maintain an insurance or other financial security for an amount equivalent to one hundred and thirty-three special drawing rights for each ton of the ship's tonnage or fourteen million elaborate special drawing rights whichever is lower.[46]

Besides the pecuniary liability for adverse environment impact the Act also charts out elaborate provision for the prevention and containment of pollution of the sea by oil. No Indian ship which has on board 2000 tons or more oil bulk as cargo shall enter or leave or attempt to enter or leave any port or place in India unless it carries a certificate[47] issued under the sub-section (2) of Section 352N or a certificate accepted under Section 352O. No Indian ship, which has on board 2000 tons or more oil in bulk as cargo shall

---

[42] *Id.* Section 352 I.
[43] *Id.* Section 352 J.
[44] *Id.* Section 352 K.
[45] *Id.* Section 352 N.
[46] *Id.* Section 352 O.
[47] *Id.* Section 352 P.

enter or leave or attempt to enter or leave any port or place in India[48] unless it carries on board a certificate issued under sub-section (2) of section 352N or a certificate accepted under section 352O. These provisions shall apply to tanker of one hundred and fifty tons gross or more and ships of five hundred tons gross or more and off-shore installation.

11.4.2 APPLICATION OF MARTIME ZONE ACT, 1976 :

Since the Act mitigate the environmental impact of sea pollution it defines sea pollution to mean any part of the territorial waters of India,[49] or any marine areas adjacent thereto over which India has, or any marine areas adjacent thereto over which India has, or may hereafter have exclusive jurisdiction in regard to control of marine pollution[50] under the territorial waters, continental shelf, (EEZ) and other Maritime Zones Act, 1976 or any other law for the time being in force.[51] In fact India's obligation to EIA of sea stems from International Convention for the Prevention of Pollution of the Sea By Oil, 1954, signed in London on the 12th day of May, 1954. The provision relating to prevention of pollution reads that no oil or oily mixture shall be discharged from an Indian tanker anywhere into the sea or from a foreign tanker anywhere within the coastal waters in India except where each of the following conditions is satisfied, namely:

---

[48] *Id.* Section 356 A.

[49] The expression "from nearest land" shall mean the baseline from which the territorial sea of the territory in question is established in accordance with the *Geneva Convention on the Territorial Sea and the Contiguous Zone*, 1958 expect that in relation to north-eastern coast of Australia it shall mean from a line drawn from a point on the coast of Australia in latitude 11 „South, longitude 142^{0}08' East to a point in latitude 10^{0}35' South, longitude 141^{0}550 East- thence to a point latitude 10^0 000 South, longitude 142^0 00' East; thence to a point latitude 9^0 10 South, longitude 143^0 52' East; thence to a point latitude 9^0 00 South, longitude 144^0 30' East; thence to a point latitude 13^0 00 South, longitude 144^0 00' East; thence to a point latitude 15^0 00 South, longitude 146^0 00' East; thence to a point latitude 18^0 00 South, longitude 147^0 00' East; thence to a point latitude 21^0 00 South, longitude 153^0 00' East; and thence to a point on the coast of Australia in latitude 24^0 42 South, longitude 253^{0}15 East.

[50] 'Oil' means – crude Oil; fuel Oil; heavy diesel oil conforming to such specifications as may be prescribed; and lubricating oil.

[51] *Id.* Section 356 B.

(a) the tanker is proceeding en-route;
(b) the instantaneous rate of discharge of oil content does not exceed sixty liters per mile;
(c) the total quantity of oil discharged does not exceed 1/15,000 part of the total carrying capacity of the tanker;
(d) the tanker is more than 50 miles from nearest land; and
(e) the tanker is not within the designated areas notified as such under sub-section (6) of Section 7 of the territorial waters, continental shelf, EEZ and other *Maritime Zones Act, 1976*.[52]

No Oil or oily mixture shall be discharged from an Indian ship other than a tanker anywhere into the sea or from a foreign ship other than a tanker within the coastal waters of India except where the following conditions is satisfied namely:

(a) the ship is proceeding en-route;
(b) the instantaneous rate of discharge of oil content does not exceed sixty liters per mile;
(c) the oil content of the discharge is less than one hundred parts per million parts of the oily mixtures;
(d) the discharge is made as far from nearest land as practicable; and
(e) the ship is not within the designated areas notified as such under sub-section (6) of Section 7 of the territorial waters, continental shelf, EEZ and other Maritime Zones Act, 1976.

The discharge of oil or oily mixture into the sea from any offshore installation is hereby-prohibited.[53] It also calls for technological renovation to prevent oil pollution into the sea and frequent auditing. The Central Government may make rule requiring Indian ships to be fitted with such equipment

---

[52] *Id.* Section 356 C.
[53] Ibid.

and to comply with such other requirements (including requirements for preventing the escape of fuel oil or crude oil or heavy diesel oil) as may be prescribed.[54] Every Indian tanker and every other Indian ship which uses oil as fuel shall maintain on board the tanker or such other ship an oil record book in the prescribed form the manner in which the oil record book shall be maintained. The nature of entries to be made therein, the time and circumstances in which such entries shall be made, the custody and disposal thereof, and all other matters relating thereto shall be such as may be prescribed having regard to the provisions of the Convention.[55] If on report from a surveyor or other persons authorized to inspect a vessel under Section 356G, the Central Government is satisfied that any provision of the Convention apply, it shall transmit particulars of the alleged contravention to the Government of the country to which the ship belongs.[56]

Under the post- EIA monitoring clauses the Act makes out a case of containment of accidental pollution. Where the central Government is satisfied that oil is escaping or is likely to escape from a tanker, a ship other than a tanker or any off-shore installation the oil so escaped or likely to escape is causing or threatens to cause pollution of any part of coasts or coastal waters of India. It may, for the purpose of minimizing the pollution already caused, or for preventing the pollution threatened to be caused require the owner, agent, master of charter of the tanker ship mobile off-shore installation, by notice served on him or as the case may be on them to take such action in relation to the tanker, ship other than a tanker, mobile off-shore installation, or as the case may be, off shore installation of any other type or its cargo or in relation to both, as may be specified in such notice. The Central

---

[54] *Id.* Section 356 E.
[55] *Id.* Section 356 F.
[56] *Id.* Section 356 H.

Government may, by any notice issued under sub-section (1) prohibit the removal-
    (a) of the tanker, ship other than a tanker, mobile off-shore installation or off-shore installation of any other type, from a place specified in the notice;
    (b) from the tanker, ship other than a tanker, mobile off-shore installation or off-shore installation of any other type of any cargo or stores as may be specified in the notice except with its previous permission and upon such conditions, if any, as may be specified in the notice.[57]

Under the imminent danger and emergency the government can proceed to take such measures as may be deemed necessary and carrying out the directives given in the notice issued under section 356J; and containing the pollution already caused or preventing the pollution threatened to be caused, of coastal waters, or, as the case may be, of any part of the coast of India by oil escaped or threatening to escape from the tanker, ship other than a tanker, a mobile off-shore installation or off-shore installation of any other type.[58] The Central Government may, having regard to the provisions of the Convention, make rules to incorporate other EIA principles for the prevention of oil pollution and protection of marine resources under intra and extra-territorial jurisdiction.[59]

11.5 HERITAGE SITES PROTECTION LAWS

The central EIA and regional EIA laws in India manifested trends of heritage conservation more vigorously under the regime of delegated legislations. These notifications owe their genesis to the *Environment Protection Act*, 1986 and *EIA Notifications* of 1994 and 1997. The study on this count revolves primarily the analysis of *Himalayan Bio-Diversity Protection Notification*, 2000 and *Taj*

---

[57] *Id.* Section 356 J.
[58] *Id.* Section 356 K.
[59] *Id.* Section 356 O.

Mahal Heritage Protection Notification, 1999 to reflect the modern trends and broad contours of heritage conservation laws under the framework of EIA law.

11.5.1 TAJ HERITAGE PROTECTION NOTICATION, 1999

In order to protect the architectural heritage of historical monument the Central Government constituted an authority to be known as *the Taj Trapezium Zone Pollution (Prevention and Control) Authority* in 1999[60] to monitor the progress of the implementation of various EIA schemes for protection of the Taj Mahal and Programmes for protection and improvement of the environment in the said area. All necessary steps are to be taken to ensure compliance of specified emission-standards by motor vehicles and ensuring compliance of fuel quality standards. The Authority shall furnish an EIA and compliance report about its activities at least once in two months to the MoEF.

11.5.2 HIMALYAN BIO-DIVERSITY PROTECTION NOTIFICATION, 2000

The *Himalayan Bio-Diversity Protection Notification*, 2000[61] has adopted the modus operandi of institutionalization and internalization of EIA law and policy is the declaration of eco-sensitive zone and heritage sites by taking into consideration the geographical and biological diversity of India. The MoEF is framing for rules the protection and improvement of Himalayan ecology and Bio-diversity. This include the States of Arunachal Pradesh, Jammu and Kashmir, Manipur, Meghalaya, Mizoram, Nagaland, Sikkim, Tripura and districts of Dehra Dun, Haridwar, Almora, Pithoragarh, Chamoli, Pauri Garhwal, Nainital, Uttar Kashi, Udham Singh Nagar, Rudra Prayag, Bageshwar, Tehri, Garhwal and Champawat of Uttar Pradesh and Darjeeling district of West Bengal.[62] In order to ensure environmentally sound

---

[60] MoEF, Government of India, Taj Trapezium Zone (Prevention and Control) Authority Notification, 1998 [S.O. (E) dated 13.5 1998].
[61] MoEF, The Draft Protection & Improvement of Himalayan Environment, 2000 [S.O. 916 (E) dated 6.10.2000].
[62] *Id.* Preamble

development and inclusion of full scale EIA cycle of hill towns, the following restrictions and conditions are proposed for all future activities in the areas in the Himalayan region.[63] It entails that no construction should be undertaken in areas having slope above 30° or areas which fall in hazard zones or areas falling on the spring lines and first order streams identified by the State Governments on the basis of available scientific evidence.[64] Construction should be permitted in areas with slope between 10° to 30° or spring recharge areas or old landslide zones with such restrictions as the competent local authority may decide.[65] Tourist resorts, commercial complexes and institutional buildings should be located in areas with surplus water and electricity so as not to affect the rights of existing users without their prior consultation.[66] Where cutting in an area causes ecological damage and slope instability in adjacent areas, such cuttings shall not be undertaken unless appropriate measures are taken to avoid such damages[67] of importance is the thrashing out of an integrated development plan taking into consideration environmental and other relevant factors including ecologically sensitive areas, hazard zones, drainage channels, steep slopes and fertile land. Areas rich in ground water may not be diverted for construction activities.[68] In order to minimize the adverse environmental impact of water resources the traditional rain water harvesting was put into effect.[69] All buildings to be constructed in future in urban areas should have provision for roof-top rain water harvesting commensurating with its plinth area with minimum capacity of 5 KL for plinth area above 200 sq.m. 2KL for plinth area of 200 sq.m. or below in case of residential buildings and minimum capacity of 0.01 cum per sq.m. of plinth area in case of commercial and institutional buildings such as tourist complexes, hotels, shopping complexes, and Government

---

[63] *Id.* Para 1.
[64] *Id.* Para 1(i).
[65] *Id.* Para 1 (ii).
[66] *Id.* Para 1 (iii).
[67] *Id.* Para 1 (iv).
[68] *Id.* Para 1 (v).
[69] *Id.* Para 2.

buildings.[70] The institutional and commercial buildings should not draw water from existing water supply schemes, which adversely affects water supply to local villages or settlements.[71] In rural areas rain water harvesting should be undertaken through such structures as percolation tanks and storage thanks and any other means. Springs sanctuary development should be undertaken in the spring recharge zones to augment spring water discharge,[72], rain water collected through storm water drains should be used to clean the waste disposal drains and sewers.[73] Groundwater aquifer recharge structures should be constructed wherever such structures do not lead to slope instabilities.[74]

The EIA dimensions of highways and roadways are also given adequate focus.[75] For construction of any road in the Himalayan region of more than 5 km (including extension/widening of existing roads) length where the same may not be tarred roads and EIA is otherwise not required. In other cases the EIA should be carried out in accordance with instructions to be issued for this purpose by the State Governments.[76] Provisions should be made in the design of the road for treatment of hill slope instabilities resulting from road cutting, cross drainage works and culverts using bio-engineering and other techniques by including the cost of such measures in the cost estimate of the proposed road.[77] Provisions should also be made for disposal of debris from construction sites in appropriate manner at suitable and identified locations so as not to affect the ecology of the area adversely. In addition to it the dumped material should be treated using bio-engineering and other appropriate techniques and the cost of such measures should be included in the cost estimate of the proposed road.[78] Wherever hot mix plants are used, they should be set up at

---

[70] *Id.* Para 2 (i).
[71] *Id.* Para 2 (ii).
[72] *Id.* Para 2 (iii).
[73] *Id.* Para 2 (iv).
[74] *Id.* Para 2 (v).
[75] *Id.* Para 2 (vi).
[76] *Id.* Para 2 (vii).
[77] *Id.* Para 3.
[78] *Id.* Para 3 (i).

least 2 km away from settlements and a minimum area of 200 sq m. surrounding the site should be devoid of vegetation.[79] No stone quarrying should be carried out without proper overall management and treatment plan including rehabilitation plan and financial provision for rehabilitation of the site should be included in the cost of the management plan.[80] All hill roads should be provided with adequate number of road side drains and these drains shall be kept free from blockage for runoff disposal. In the event that this is not done and this fact leads to damages that could otherwise have been prevented. The persons responsible should be liable for prosecution/damages, further, the cross drains shall be treated suitably using bio-engineering and other appropriate technologies so as to minimize slope instability.[81] The run-off from the road side drains should be connected with the natural drainage system in the area.[82] Fault zones and historically land slide prone zones should be avoided during alignment of a road, where for any reason it is not possible to do so,[83] notice should be given providing full justification and the construction should be carried out only after sufficient measures have been taken to minimize the associated risks.[84] Notice should be given about all fault zone and land slide zone along the roads indicating the beginning and the end of such area[85] alignment should be preferred to valley alignment[86]; to minimize loss of vegetal cover[87] and south or south-west alignment should be preferred to avoid moist areas.[88] Appropriate design standards should be followed while designing the roads including mass balancing of cut and fill and avoidance of

---

[79] *Id.* Para 3 (ii).
[80] *Id.* Para 3 (iii).
[81] *Id.* Para 3 (iv).
[82] *Id.* Para 3 (v).
[83] *Id.* Para 3 (vi).
[84] *Id.* Para 3 (vii).
[85] *Id.* Para 3 (viii).
[86] *Id.* Para 3 (ix).
[87] *Id.* Para 3 (x).
[88] *Id.* Para 3 (xi).

unnecessary cutting[89] and encouragement should be provided for use of debris material for local development.[90]

In quick succession of issuance of EIA Draft Notification For Protection And Improvement of Quality of Environment in the Himalyan Regions, the MoEF proposed to issue a Notification To Protect and Improve the Quality of Environment in the Himalayas,[91] to ensure environmentally sound development and full scale inclusion of EIA in Himalyan regions including States of Assam and Himachal Pardesh in 2000.[92]

11.6 CONCLUSION AND SUMMATIONS

An analysis of the conventions and recommendations reveals that the international law governing the heritage conservation has been a case of delayed attention. the growth of the law on the subject is marked by the specific term of reference to the preservation of cultural heritage than that of man-made and natural heritage .These laws having narrow objectives being myopic to the objective and limited subject matter directed towards the war like situations and armed cionflicts. Despite these shortcomings, the legislative initiatives of international law and UNESCO reflect the environmental concerns in rudimentary form embodying an approach to protection of cultural property as "nationalist" or "statist" ideals of the state. This also implies that the interest of the State of origin which often happens to be the developing world should be paramount. The modern heritage conservation laws also reflected both the growing concern in environmental issues in its integration of the cultural with the natural heritage as well as the concept of a "common heritage of mankind". This seem as a quantum leap and the national legislatures should take mileage in moulding the environmental law, environment impact assessment(EIA) and land laws and heritage conservation laws.

---

[89] *Id.* Para 3 (xii).
[90] *Id.* Para 3 (xiii).
[91] MoEF, The Draft Notification on Protection and Improvement of Himalyan Environment In Assam And Himachal Pradesh, 2000 [S.O. 1058 (E) dated 28.11.2000].
[92] *Id.* Preamble.

Quite significant is the imperial statutory attempt of acquisition of land to pursue the colonial interests and indigenous development. Despite the colonial bias the independent India without much of the radical structuring of the Act, implemented widely even at the cost of adverse social and environmental impacts. The blanket power of acquisition of land under the euphemistic phrase of national interest and development most often than not led to widespread unsustainable maneuvering land resources of the country. It bears enough testimony that in the name development such acquisitions, subsidized the life style of poor and impoverished them to great social and ecological imbalance. It was only recently that the damaging portent of the law was seriously called into question and there is gradual attempt if refinement in the light of EIA principles. The hidden potentiality of heritage conservation legislation such as *Ancient Monument Preservation Act,* 1904 and *Ancient Monuments and Archeological Sites Remains Act,* 1958 in reference to EIA is rarely utilized though the law exhibits a resilience to cope with the EIA imperatives the fortuitous dimensions of the provisions still needs to be harnessed. The EIA of off-shore and coastal regulation zone is impressively displayed under *Merchant Shipping Act,* 1958. While specifically dealing with marine pollution under the territorial water, continental shelf, exclusive economic zone and maritime zone the Act cast responsibilities for undertaking EIA process by the navigation and ship industries. Nowhere the Trans boundary dimensions of EIA find a categorical mention under any Indian law as it seems explicit in the *Merchant Shipping Act,* 1958.

REFERENCE

PRIMARY SOURCE

1. Preamble to UNESCO,,s Convention on Protection of Cultural Property in the event of Armed Conflict, 1954
2. The Land Acquisition Act, 1884 [Act of 1894]: The text of the Act reflects the law as on 31.3.1996.
3. The Land Acquisition Amendment Act, 1962
4. The Land Acquisition Amendment Act, 1967 [The Act 13 of 1967]

5. The Ancient Monument Preservation Act, 1904 [The Act VII of 1904]: The text of the Act reflects the law as on 15.4.1969.
6. The Ancient Monument Preservation (Amendment) Act, 1932.
7. The Ancient Monuments and Archaeological Sites and Remains Act, 1958 [The Act of the XXIV of 1958]: The text reflects the law as on 15.4.1969.
8. The Statement of Objects And Reason, Merchant Shipping Act, 1958 Para 1.
9. Secondary Sources
10. The Recommendation on International Principles Applicable to Archeological Excavation, 1956 [UNESCO 5dec.1956]
11. Recommendation Concerning The Most Effective Means Of Rendering Museum Accessible To Everyone, 1960[UNESCO 14 Dec.1960]
12. Recommendation Concerning The Safeguarding Of The Beauty And Character Of Landscape And Sites,1962 [UNESCO 11 Dec.1962]
13. Recommendation Concerning The Preservation Of Cultural Property Endangered By Public Work Or Private Work,1968 [UNESCO 19 Nov.1968].
14. Convention Relating To The Prohibition and Prevention of Trafficking in Cultural Property, 1970
15. Convention on The Means of Prohibiting and Preventing The Illicit Import, Export and Transfer of Ownership of Cultural Property,1970 [14 Nov 1970[823U.N.T.S.231]
16. Convention Concerning the Protection of the World Cultural and Natural Heritage,1972[16 Nov.1972,11 I.L.M.1358]]
17. Janet Black, International & Comparative Law Quarterly, „On Defining The Cultural Heritage', 49(1),61(2000)
18. MoEF, Government of India, Taj Trapezium Zone (Prevention and Control) Authority Notification, 1998 [S.O. (E) dated 13.5 1998].
19. MoEF, The Draft Protection & Improvement of Himalyan Environment, 2000 [S.O. 916 (E) dated 6.10.2000].
20. MoEF, The Draft Notification on Protection and Improvement of Himalyan Environment in Assam and Himachal Pradesh, 2000 [S.O. 1058 (E) dated 28.11.2000].

Chapter 12

Environmental Impact Assessment in India: An Analysis of Law and Judicial Trends in Contemporary Perspective

Anis Ahmad

Abstract: *Environmental impact assessment (EIA) has evolved and become part of major developmental project requirements in many countries including India. The significance and relevance of EIA are often subjected to congenialities of economic development and ecological values. The methodology adopted is that of self-assessment by the project proponent followed by review and project approval by the regulators created by law. This is more apt in Indian context because it is one of the largest democratic country, rich bio-diversity and natural resources, traditional environmentalism and bureaucratic intervention of the state sponsored developmental projects regions of the world. This necessitates a vibrant EIA law and policy to realise the basic precepts of sustainable development. However, it is worth noting that environmental management is not achieved only through environmental laws but also through the application of various formal and informal administrative mechanisms. EIA was first introduced as regulatory requirement only in 1994 and amended in 2006 to balance economic development and environment protection to minimize the impact. The existing ontology of EIA law in India is glaringly conspicuous by its gross absence of socio-legal researches oriented to fill void of policy, law enforcement and development. With the background, the paper seek to examine the development of EIA law and policy and its translate into reality by the Supreme Court as well as the National Green Tribunal (NGT)as green institutional have developed a rich* *environmental jurisprudence in recent past from 2000-2017 in order to consider the importance of EIA process for achieving the sustainable development goals in coming future through their judicial decisions.*

Keywords: *Environmental impact assessment, sustainable development, Supreme Court, National Green Tribunal.*

12.1 INTRODUCTION

The linkage between environmental management and economic development becomes an important discourse for policy makers at international and national level over the past few decades. (Birnie, Boye & Redgwell,2009) The quest to protect the environment from further degradation has been become global agenda for the survival and well-being of human on this earth for many years now. The eradication of poverty, population growth, natural resources depletion, water pollution and sanitation, inequality, human settlement, food insecurity still exist as biggest challenge in order to realize the sustainable development goals (SDGs) in coming future 2030. (Kothari, 2002) Keeping in view the international and national legal regime engrafted environmental impact assessment (EIA) procedures as *sine quo non* for sustainable development in compliance with the Principle 17 of the UN Conference on Environment and Development in 1992. (Nomani, 2010) However, it is worth mentioning that environmental management is not achieved merely by making environmental laws and policies but it also requires by applying various formal and informal administrative mechanisms such EIA procedure which was adopted by the US National Environmental Policy Act (NEPA) in 1969. (Betey& Godfred, 2013) Further, the EIA as an instrument of environmental law has not been given sufficient attention because it is only a kind of procedural requirement in the proposed developmental projects in beginning in India. Prior to the promulgation of EIA *Notification of* 1994, EIA has been ingrained in discretionary model under administrative framework for environmental impact assessment is unsuited for integrating economic and ecological considerations in decision making. Of late only in 1994 the first EIA notification was issued by the Central Government in exercise of its power to take any measure to protect and improve the environment as provided under Section 3 of *the Environment Protection Act*, 1986. Thereafter significant changes were initiated through amended version of *Environment Impact Assessment Notification*, 2006 that superseded the 1994 notification. (Chowdhary, 2016) The review of the significant case laws related to EIA process is also worth mentioning to develop a rich environmental jurisprudence from time to time by Supreme Court and particularly the role of National Green Tribunal commendable from its inception. Under this background the present paper wishes to analyses the EIA Law and judicial trends in last couple of decades and more in order to realize the sustainable development in coming future.

12.2 CONCEPTUAL FRAMEWORK OF EIA

In order to understand the concept of EIA, one first needs to know what an „Environmental Impact' is. An „Environment Impact' is any impact or effect (positive or negative) that an activity has on an environmental system, environmental quality or natural resources, it is also known as an Environmental effect (Oxford Dictionary of Environment and Conservation, first Edn., 2007). An „Environmental Effect' is defined as a natural or artificial disturbance of the physical, chemical or biological components that make up Environment (Black's Law Dictionary, 9th Edn., 2009). Such activities may take the form of mining, oil and gas exploration, thermal, nuclear and hydraulic power plants, metallurgical industries, chemincal estates/parks/complexes/areas, waste treatment plants, etc.

The term „environmental impact assessment' is used broadly to include a whole range of social and economic impacts. No matter how the terms are used, it is important to recognize that impacts on ecosystems, biogeochemical cycles, and the like are intimately related through complex feedback mechanisms to social impacts and economic considerations.

Environment Impact Assessment (EIA) can be defined as the study to predict the effect of a proposed activity/project on the environment and natural resources. A decision making tool, EIA compares various alternatives for a project and seeks to identify the one which represents the best combination of economic and environmental costs and benefits. (Bhushan &Kumar, 2006),

EIA systematically examines both beneficial and adverse consequences of the project and ensures that these effects are taken into account during project design. It helps to identify possible environmental effects of the proposed project, proposes measures to mitigate adverse effects and predicts whether there will be significant adverse environmental effects, even after the mitigation is implemented. By considering the environmental effects of the project and their mitigation early in the project planning cycle, environmental assessment has many benefits, such as protection of environment, optimum utilisation of resources and saving of time and cost of the project.

In other word the Environmental impact assessment, which is defined as an activity designed to identify and predict the impact on the bio geophysical environment and on man's health and well-being of legislative proposals, policies, programmes, projects, and operational procedures, and to interpret and communicate information about the impacts of that activities.

Environmental Impact Assessment (EIA) is also a kind of effort to anticipate measure and weigh the socio-economic and bio-physical changes that may result from a proposed project. It assists decision-makers in considering the proposed project's environmental costs and benefits. Where the benefits sufficiently exceed the costs, the project can be viewed as environmentally justified. In view of the intricate web of relationships between the different parts of an ecosystem, a comprehensive EIA inevitable requires a multi-disciplinary approach. An environmental impact statement (EIS) for a dam, for example, might include inputs from geologists, forestry experts, wildlife experts, anthropologists, economists, agricultural scientists and social scientists. (Rosencranz & Divan, 2001)

The primary aim of EIA procedures is to gauge the potential environmental impact of an economic project so as to allow for measures to minimize that impact. The methodology adopted is that of self-assessment by the project proponent followed by review and project approval by the regulators. It is essential that consequences of projects, plans or policies at different levels be assessed before they are executed. EIA examines these consequences and predictions for future changes in the environment. It guides administrative agencies in balancing conflicting social values and environmental quality.(Leelakrishnan,1999)

The requirement of a state to conduct an environmental impact assessment (EIA) in respect of activities with the potential to significantly affect the environment is well reflected in Principle 17 of the UN Conference on Environment and Development (UNCED) (United Nations, 1992). "Environmental impact assessment, as a national instrument, shall be undertaken for proposed activities that are likely to have a significant adverse impact on the environment and are subject to a decision of a competent national authority". The UNCED recognized EIA as a key tool for environmental protection and sustainable development. By implication, unless sustainable development criteria are included specifically among those uses in environmental assessment, EIA may not contribute to sustainable development.

The evolution of EIA is one of the successful policy innovations of the 20th Century for environmental protection. Nearly five decades ago, there was no EIA process for any developmental projects but today, it is a formal process in many countries of the world and is currently practiced in more than 100 countries. EIA as a mandatory regulatory procedure originated in the early 1970s, with the implementation of the National Environment Policy Act (NEPA) 1969 in the US. A large part of the initial development took place in a few high-income countries, like Canada, Australia, and New Zealand (1973-74). However, there were some developing countries as well, which introduced EIA relatively early Columbia (1974), Philippines (1978) and India (1994).

The EIA process really took off after the mid-1980s. In 1989, the World Bank adopted EIA for major development projects, in which a borrower country had to undertake an EIA under its supervision. Thus, an EIA in general parlance does not confine itself only to projects but also to legislations and policies.

12.3 EIA LAW AND POLICY

The evolutionary phase of EIA law in India discerns a trend from conventionalism to legalism. In the pre-1970 phase the internalization of EIA principles found a specific mention under the term of reference of National Committee of Environment Planning and Co-ordination (NCEPC) established in Department of Science and Technology (DST). During, 1970s the Indian Constitution by (Forty Second Amendment Act), 1976 incorporated environment protection measures enjoining state as well as citizens to protect and improve the natural environment. As a natural corollary Fifth National Plan stressed close association of NCEPC with major industrial decisions and placed priority to minimize pollution. During 1980-85 recognition of central authority for environmental protection and conservation was envisioned. A new mechanism of environment protection included Planning Commission's approval of major projects and review report by Department of Environment (DoE). Environmental appraisal and monitoring committees were set up to review impact statements by project proponents. Guidelines and checklists for EIA were developed for hydroelectric, mining, harbor, thermal and road and rail projects. Environmental boards in all states and Union territories established for review and monitoring of all environmental matters. In order to protect federal principle of environmental management and EIA the Ministry of Environment and Forest (MoEF) has issued series of notifications relating protection of ecologically fragile and sensitive zones of country. These notifications empowered federal identities to actualize the basic precepts of sustainable development and EIA. (Nomani, 2010)

Prior to January 1994, EIA in India was carried out under administrative guidelines which required the project proponents of major irrigation projects, river valley projects, power stations, ports and harbours, etc., to secure a clearance from the Union Ministry of Environment and Forests (MEF). The procedure required the project authority to submit environmental information to the MEF by filling out questionnaires or checklists. The environmental appraisal was carried out by the ministry's environmental appraisal committees. These committees held discussions with the project authority and on the basis of the deliberations either approved or rejected the site.

When approved, the project clearance was generally made conditional on specified safeguards.

The EIA notification was first issued in 1994 by the Central Government (Ministry of Environment and Forests (MOEF)) in exercise of its power to take any measure to protect and improve the environment as provided under Section 3 of *the Environment Protection Act, 1986*. It introduced a process for prior environmental approval of certain kind of projects (specified in Schedule). The project proponents were required to submit an environmental assessment report, environmental management plan and the details of the public hearing conducted in the vicinity of the project (exceptions to these requirements were permitted for certain projects). The MOEF would function as Impact Assessment Agency which could consult a Committee of Experts set up for this purpose. On 27 January 1994, the MEF notified mandatory EIAs under Rule 5 of the Environment (Protection) Rules of 1986 for 29 designated projects. The notification made it obligatory to prepare and submit an EIA, an Environment Management Plan (EMP), and a Project Report to an Impact Assessment Agency for clearance. The MEF was designated as the Impact Assessment Agency and was required to consult a multi-disciplinary committee of experts. Under the January 1994 notification any member of the public was to have access to a summary of the Project Report and the detailed EMPs. Public hearings were mandatory. This represented India's first attempt at a comprehensive EIA scheme. On 4 May 1994 the MEF issued an amending notification substantially diluting the January 27 notification, On 10 April 1997 some of the regressive changes introduced in May 1994 were undone by fresh amendments to the parent notification. The 1997 provisions restore public hearings, to be conducted in the manner prescribed. On the same day the MEF published a separate notification prescribing the EIA procedure for clearance of certain types of thermal power plants requiring environmental clearance from the concerned state government.

Three significant changes were initiated through the 2006 amendment that superseded the 1994 notification. The Ministry of Environment and Forest (MoEF) introduced an amended version of *Environment Impact Assessment Notification,* 2006. The law came to statue book under the influence of *Govind Rajan Committee Report on Investment Reform, Environment Management Capacity Building Project of World Bank and New Economic Policy.* Thus the Notification clearly prioritizes the needs of industry and investment over environmental and social concerns stifling the professed goal of public consultation and sustainable development. It has been experienced the EIA is an important management tool for integrating environmental concerns in development process and for improved decision making necessitates radical changes in view of liberalization, privatization and globalization, to achieve this, the Ministry had notified certain developmental activities which could be taken up only after prior environmental clearance from the Ministry under environmental regulations such as EIA Notification and Coastal Resource Zone (CRZ) Notification, Environmental clearances based on EIA Study was through EIA Notification, 1994 under the *Environment (Protection) Act,* 1986. Over the period, certain constraints were experienced in smooth implementation of the Notification. Therefore, Ministry reviewed comprehensively the existing environmental clearance process for further enhancing the quality of the appraisal and to reduce time in the decision-making within the prescribed statutory period. After holding extensive consultations of all the suggestions received, the Ministry notified the *EIA Notification,* 2006 on September 14, 2006 superseding the EIA Notification, 1994. With the Notification, environmental clearance process has been re-engineered. The ministry maintains that:

The notification is a path breaking re-engineered process and is comparable to the best practices followed internationally. The important changes include introduction of screening and scoping of the project proposals for the identification of the actual environmental priorities without asking for irrelevant and time consuming studies.

Under the new Notification, the projects will now require prior environmental clearance based on the impact potential instead of investment criteria. Public Hearing proceeding are also better structured and time bound. The environmental clearance process has also been decentralized for certain categories of projects, termed as Category „B' projects, which are below a prescribed threshold level. Such projects will be appraised at State level by constituting a State

Level Environment Impact Assessment Authority (SEIAA) and State Level Expert Appraisal Committee (SEAC). These will be constituted by the Central Government in consultation with the State Governments / UTS Administration. For construction projects, considerable simplification in the application with appraisal procedures has been proposed, so that development of this sector is not retarded'. Such projects have been exempted from the public hearing also. The notification also provides for exemption from public hearing for certain categories/types of projects.

In post 1994 phase the National Environment Policy 2006 seeks to extend the coverage and fill in gaps that still exist in light of present knowledge and accumulated experience. The present national policies for environmental management are contained in National Forest Policy,1988, the National Conservation Strategy and Policy Statement on Environment and development,1992, and Policy Statement on Abatement of Pollution,1992, the National Agriculture policy, 2000, the National Water Policy, 2012 having broad sectoral coverage of EIA principles in their contents for sustainable development. But due to space limit I will discuss only the first comprehensive environmental policy i.e. the National Environment Policy 2006(NEP, 2006).

One of the principal objectives of this policy is enumerated below:

"To integrate environmental concerns into policies, plans, programmes, and projects for economic and social development."

Apart from that the Policy has enough potentiality and having broad coverage of EIA principles within the framework of sustainable development. The incorporation of EIA Principles under the NEP 2006 well reflected in the following paragraphs of the policy:

Environmental Impact Assessment (EIA) will continue to be the principal methodology for appraising and reviewing new projects. The assessment processes are under major revision in line with the Govindrajan Committee recommendations. Under the new arrangements, there would be significant devolution of powers to the State/UT level. However, such devolution, to be effective, needs to be accompanied by adequate development of human and institutional capacities.

Further, in order to make the clearance processes more effective, the following actions will be taken:

a. Encourage regulatory authorities, Central and State, to institutionalize regional and cumulative environmental impact assessments (R/CEIAs) to ensure that environmental concerns are identified and addressed at the planning stage itself.
b. Specifically assess the potential for chemical accidents of relevant projects as part of the environmental appraisal process.
c. Give due consideration, to the quality and productivity of lands which are proposed to be converted for development activities, as part of the environmental clearance process. Projects involving large-scale diversion of prime agricultural land would require environmental appraisal.
d. Encourage clustering of industries and other development activities to facilitate setting up of environmental management infrastructure, as well as monitoring and enforcing environmental compliance. Emphasize post-project monitoring and implementation of environmental management plans through participatory processes, involving adequately empowered relevant levels of government, industry, and the potentially impacted community.
e. Restrict the diversion of dense natural forests and areas of high endemism of genetic resources, to non-forest purposes, only to site-specific cases of vital national interest. No further regularisation of encroachment on forests should be permitted.
f. Ensure that in all cases of diversion of forest, the essential minimum needed for the project or activity is diverted. The diverted area must not be cleared until the actual construction starts.
g. Ensure provision for environmental restoration after decommissioning of industries, in particular mine closure in all approvals of mining plans, and institutionalize a system of post-monitoring of such projects.
h. Formulate, and periodically update, codes of "good practices" for environmental management for different categories of regulated activities.

Further, the statement of policy Objectives and Principles are to be realized by concrete strategies and actions in different areas relating to key environmental challenges.

12.4 JURISTIC ARTICULATION OF EIA

The juristic articulation related to EIA in India is a case of mixed and varied responses. The series of judicial pronouncements in matters environment protection has evolved and developed environmental jurisprudence by focusing on EIA principles in pre and post EIA Notification, 1994 phase in order to provide environmental justice. The contemporary judicial trend is the study of cases decided by the Supreme Court since 2000 and thereafter as well as the National Green Tribunal after its inception in 2010 expanded upon and deepened the impact of these changes through their decisions which developed key aspects of the EIA process.

12.4.1 Supreme Court's Jurisprudence on EIA

In *Narmada Bachao Andolan v Union of India* (2000), the Supreme Court was confronted with a case in which a massive developmental project (Sardar Sarovar Project dam on the river Narmada) was challenged on the ground of non-completion of the environmental impact assessment (EIA) and the inadequate rehabilitation and resettlement efforts made for the project-affected persons. The petitioners and argued that this was a fit case for the application of the precautionary principle since the dam would potentially cause irreparable damage to the local environment. The Curt sought to address this aspect by making a specious argument that this principle is applicable in case where there is uncertainty prevailing as to the extent of damage or pollution. However, this was not such as case, because the effect of the dam on the environment was known. This was not completely true since EIA was not under-taken for the entire project. So at best the Court was making a claim based on the impact of dams in general as per anecdotal evidence. The court went on to state that:

> "Merely because there will be a change is no reason to presume that there will be ecological disaster. It is when the effect of the project is known then the

> principle of sustainable development would come into play which will ensure that mitigative steps are and can be taken to preserve the ecological balance. Sustainable development means what type or extent of development can take place which can be sustained by nature/ecology with or without mitigation."

In *Centre for Social Justice v Union of India* (2001) is a case of seminal importance is giving effect to the public participation under EIA laws in latter and spritis. Since this is the first case dealing with the compliance of public participation public hearing provision in EIA law in the country.

In *ND Jayal and Another v Union of India and Others* (2004) was a similar case in which the Tehri Dam project was challenged on environmental safety issues. The Supreme Court adopted a similar position to that in the Narmada Bachao Andolan case. On the face of obvious non-compliance of the project proponent, it chose to focus on the economic gains from the project (in this case a dam to generate hydroelectric power).

Interestingly, even the dissenting opinion of Justice Dharmadhikari in this case reflects the thinking of the majority decision in terms of the way in which environment implications can and will necessarily be addressed while going ahead with development decisions. He observed that:

> "A strategy for conserving or resources-effective use of non-renewable resources is the imperative demand of modern times. Whereas, minimum sustainable development must not endanger the natural system that supports life on earth, constant technological efforts are demanded for resources-effective production, so that sacrifice of one eco-system is counter balanced or compensated by recreating another system."

In *Utkarsh Mandal v Union of India* (2009) the Delhi High Court had held that the EAC was bound to disclose the reasons underlying its decision following the principle enunciated by the Supreme Court that quasi-judicial and administrative bodies have to disclose reasons for reaching a particular conclusion. Further the Court has emphasized the need for a detailed analysis of facts and reasoning.

Again in Lafarge Umiam Minig Private Limited v Union of India & Ors (2011) It was observed by this Court:

> the public consultation or public hearing as it is commonly known, is a mandatory requirement of the Environment Clearance process and provides an effective forum for any person aggrieved by any aspect o0f any project to register and seek redressal of his/her grievances.
>
> ...... It cannot be gainsaid that utilization of the environment and its natural resources has to be in a way that is consistent with principles of sustainable development and intergenerational equity, but balancing of these equities may entail policy choices. In the circumstances, barring exceptions, decisions relating to utilization of natural resources have to be tested on the anvil of the well-recognised principles of judicial review. Have all the relevant factors been taken into account? Have any extraneous factors influenced the decision? Is the decision strictly in accordance with the legislative policy underlying the law (if any that governs the field? Is the decision consistent with the principles of sustainable development in the sense that has the decision-maker taken into account the said principle and, on the basis of relevant considerations, arrived at a balanced decision? Thus, the Court should review the decision-making process to ensure that the decision of MoEF is fair and fully informed, based on the co0rrect principles, and free from any bias or restraint.

Thus in Gram Panchayat Navlakh Umbre v Union of India and Ors, (2012) the Court held that the:

> "decision making process of those authorities besides being transparent must result in a reasoned conclusion which is reflective of a due application of mind to the diverse concerns arising from a project such as the present. The mere fact that a body is comprised of experts is not sufficient a safeguard to ensure that the conclusion of its deliberations is just and proper."

In *Deepak Kumar v State of Haryana and Ors,* (2012) referring to the recommendations of the Committee on Minor Minerals, the court underlined that state governments should be discouraged from granting a mining license/lease to plots less than five hectares so as to reduce circumvention and ensure sustainable mining. Further, where land is broken up into smaller parcels, prior environmental approvals should be sought from the MOEF. The court finds that:

> It is without conducting any study on the possible environmental impact on/in the river beds and elsewhere the auction notices have been issued. We ore of the considered view that when we are faced which a situation sheer extraction of alluvial material within or near a river bed has an impact on the rivers physical habitat characteristics, like river stability, flood risk, environmental degradation, loss of habitat, decline in biodiversity, it is not an answer to say that the extraction is in blocks of less than 5 hectares, separated by 1 kilometer, because their collective impact may be significant, it is not an answer to say that the extraction is in blocks of less than 5 hectares, separated by 1 kilometer, because their collective impact may be significant, hence the necessity of a proper environmental assessment plan. Possibly this may be reason that in the

affidavit filed by the MoEF on 23.11.2011 along with the annexure-2 report, the following stand has been taken.

In *Sterlite Industries (India) Ltd. v. Union of India* (2013) the Supreme Court discussed the specific grounds on which administrative action involving the grant of environmental approval could be challenged. The grounds for judicial review were illegality, irrationality and procedural impropriety. Thus the granting of environmental approval by the competent authority outside the powers given to the authority by law, would be grounds for illegality. If the decision were to suffer from unreasonableness, the Court could interfere on grounds of irrationality. Last, an approval can be challenged on the grounds that it has been granted in breach of proper procedure. Nevertheless the Court has not restrained itself, in cases where it found that the SEAC had recommended approvals without any application of Environmental Impact Assessment in India.

In *G Sundarrajan v Union of India and Ors* (2013) wherein the issue was operation of the Kudankulam Nuclear Power Project (KNPP) that was challenged on the ground of Non-completion of the EIA. The court stated that:

> Court has emphasized on striking a balance between the ecology and environment on one hand, and the projects utility on the other. The trend of authorities is that a delicate balance has to be struck between the ecological impact and development. The other principle that has been ingrained is that if a project is beneficial for the larger public, inconvenience to smaller number of people is to be accepted. It has to be respectfully accepted as a proposition of Law that individual interest or, for that matter smaller public interest must yield to the larger public interest.

The court in G Sundarrajan v Union of India and Ors (2013)

Again in *T.N Godavarman v Union of India* (2014) It was held that the:

> "present mechanism under the EIA Notification ... is deficient in many respects and what is required is a Regulator at the national level ....which can carry out an independent, objective and transparent appraisal and approval of the projects for environmental clearances and which can also monitor the implementation of the conditions laid down in the Environmental Clearances."

In *Talaulicar & Sons P. Ltd. v The Union of India* (UOI) and Ors. (2016) the court passed an order to issue a notice to the first Respondent MoEF for hearing to the Appellant and hold consultative process with State Level Authorities and call for the required reports from the concerned experts of its choice and after due hearing, pass and appropriate oders, in accordance with law.

In Electrotherm (India) Ltd. v Patel Vipul kumar Ramjibhai and Ors (2016) The court observed that:

> Public consultation/public hearing is one of the important stages while considering the matter for grant of Environmental Clearance. The minutes of the meetings held on 9-11 February, 2009 show that the request of the appellant for exemption from the requirement of public hearing was accepted by the Committee. The observations of the Committee suggest that there would be no additional land requirement, ground water drawl and certain other features. However the water requirement, which is a community resource, was definitely going to be of greater order in addition to the fact that the expansion of the project would have entailed additional pollution load.

Recently, in *Common Cause and Ors. v Union of India and Ors* (2107) the Supreme Court speaking through Justice Madan B. Lokur in the following language:

> There is no doubt that the grant of an EC cannot be taken as a mechanical exercise. It can only be granted after due diligence and reasonable care since damage to the environment can have a long term impact. EIA 1994 is therefore very clear that if expansion or modernization of any mining activity exceeds the existing pollution load, a prior EC is necessary and as already held by this court in M.C. Mehta even for the renewal of a mining lease where there is no expansion or modernization of any activity, a prior EC is necessary. Such importance having been given to an EC, the grant of an ex post facto environmental clearance would be detrimental to the environment and could lead to irreparable degradation of the environment. The concept of an *ex post facto* or a retrospective EC is completely alien to environmental jurisprudence including EIA 1994 and EIA 2006. We make it clear that an EC will come into force not earlier than the date of its grant.

12.4.2 EIA and Role of National Green Tribunal

The National Green Tribunal (NGT) has played very vital role from very inception of its establishment. The NGT's case list has been dominated by EIA matters and environmental clearances in most of the cases. Therefore, it is quite important to analyses the jurisprudential development in following EIA cases:

In *Adivasi Majdoor kisan Ekta Sangathan and Another v Ministry of Environment and Forest and Others* (2012) The Tribunal finds that the evidence of persons who voiced their opposition to the project was not recorded and no summary of the public hearing was prepared in the local language nor was it made public. Therefore the Tribunal declared the approval was invalid.

Again in *Antarsingh Patel and Ors v Union of India* (2012) in which the Maheshwar Hydro Power Project was challenged by project-affected persons. The NGT attempted to secure better legal cover for the rights of the project-affected persons by stating that:

> It is no longer res integra that the benefits of developmental activities must go to the local people and their quality of life must improve instead of driving them to a disadvantageous position. Depriving them of the facilities which they were already enjoying, but are likely to be deprived of due to the proposed Hydro Electric project would be contrary to the law. Citizens are at the centre of development and as such all efforts are required to be made to avoid any hardships to the affected persons.
>
> (Antarsingh Patel and Ors v Union of India (2012),

In Gau Raxa Hitraxak Manch and Gaucher Paryavaran Bachav Trust, Rajula v Union of India and Others (2013) the National Green Tribunal (NGT) has held that:

> "the appraisal is not a mere formality and it requires detailed scrutiny by EAC and SEAC of the application as well as the documents filed, the final decision for either rejecting or granting an EC vests with the Regulatory Authority concerned viz., SEIAA or MOEF, but the task of appraisal is vested with EAC/SEAC and not with the regulatory authority."

In Samata and Forum of Sustainable Development v Union of India & Ors (2013) the NGT held that:

> "In order to demonstrate [the] threadbare nature of discussions while considering a project for giving its recommendation, it is essential that the views, opinions, comments and suggestions made by each and every member of the committee are recorded in a structured manifest/ format."

In *M P Patil v Union of India* (2014) is yet another interesting case where the NGT pushed the envelope further even while partially dismissing the challenge. The case involved not only ecological risk, but also great social impact sine the project-affected persons—the group to be resettled and rehabilitated (R&R) were particularly large. First, while reiterating the need to balance between environment and development as is entailed by the principle of sustainable development, it stated that given the considerable impact of the project on human displacement, the R&R scheme would be one of the most pertinent aspects to be considered by the EAC. Following from this the R&R scheme had to be elaborately deliberated upon by the project proponent and considered by the EAC and the views of the general public should be heard on this issue specifically during the public hearing.

In *T Muruganandam and Ors v Union of India and Ors* (2014) Judgment of NGT, Principal Bench, New Delhi on November (2014) the environmental clearance granted by the Ministry of Environment, Forest and Climate Change (MoEF) to the Tamil Nadu Power Company was challenged. It was challenged on the ground that the cumulative environment impact assessment (CEIA) undertaken was not adhering to universally accepted scientific parameters, and therefore, bad in law. This was contested on the ground that under the Indian environment legislation scenario there are no known "universally accepted scientific parameters" for (CEIA) study. First the NGT categorically found that CEIA was required as per the precautionary principle and sustainable development.

In *S.P. Muthuraman and Ors v Union of India and Ors.* (2015) the NGT elaborated on its jurisdiction and power and stated that:

> even if the structures of the project proponents are to be protected and no harsh direction are passed in

that behalf, still the Tribunal would be required to pass appropriate directions to prevent further damage to the environment on the one hand and control the already caused degradation and destruction of the environment and ecology by these projects on the other hand. Furthermore, they cannot escape the liability of having flouted the law by raising substantial construction without obtaining prior environmental clearance as well as by flouting the directions issued by the authorities from time to time. The penalties can be imposed for such disobedience or noncompliance. The authorities have already initiated an action against three of the project proponents and have taken proceedings in the court of competent jurisdiction under Act of 1986. However no action has been taken against other four project proponents as of now. Penalties can be imposed for violation in due course upon full trial.

In *Vikrant Tongad Vs. Noida Metro Rail Corporation and Ors,* (2016) the NGT speaking through Justice Swatanter Kumar and lamented that:

"The purpose of the Notification of 2006 is not to prohibit development but to permit the same while protecting the environment and ecology. It is the requirement that there should not be irretrievable or irreversible damage to the nature and environment. In the event the project commenced damage then the entire project would fall beyond the known dimensions of principle of Sustainable Development and indiscriminate development would certainly have adverse impacts upon the environment and ecology of the area. Taking environmental clearance would cause no prejudice to any of the stakeholders on the one hand, while on the other it will protect the environment, nature, the river and its banks. The official respondents and the project proponents both had been ad idem that the project did not require prior environmental Clearance in terms of Notification of 2006. Since we have now held that the project is covered project proponent to take environmental clearance."

Recently, in *Santhiyagu Vs. The Union of India and Ors.* (2017) the NGT Southern Zone Bench orders that:

"Therefore, it is clear that when once it is concluded that the treatment involves not just domestic sewage but also industrial and trade effluents, it no longer involves setting up of STP simplicitor but it is more or less a CETP which requires prior EC. There are different parameters and inlet and outlet standards for STP and CETPs and there is a change in design, type of machinery and equipment also taking into account the toxicity and obnoxious nature of the effluent to be treated."

12.5 CONCLUSION & ASSESSMENT

An Analysis of the evolution and development of EIA in India reveals that the law and policies were passed under the broader object of „Protection and Improvement' envisioned under the constitutional precepts and legislative objects *of Environment (Protection) Act*, 19986. But the reformative journey of EIA law from 1994 to 2006 is also marred with blurred vision of sustainable development by our leading hights of law and administration. Most often than not the amendment are brought in for political expediency than the ecological necessity and viability. The judicial scrutiny in post EIA Notification phase since *Narmada case* (2000) judgment particularly the dissenting opinion of Justice S.P.Bharucha and others chisel out more dimensions to be added to EIA Law in India. A selected study of judgments of Supreme Court and the orders of the National Green Tribunal related to EIA depicts a serious view taken by these two judicial institutions to enforce the basic norms from time to time. Further, it is also important to mention that the EIA process is India faces several critical challenges, the primary being the need for greater transparency, ensuring accountability of regulators and improving the quality of public participation. The Court's interventions in various cases have sought to address each of these challenges. In spite of these challenges the role of the Supreme Court as well as NGT in order to make EIA law and policy more effective for sustainable development is welcome step. Therefore, it is high time to ponder into the EIA legal Strategies and reform need to be done in light of juristic articulation for achieving the goals of sustainable development in coming future.

REFERENCES

1. Armin Divan Shyam Rosencranz. (2001) *Environmental Law and Policy in India Cases, materials and Statutes* Second Edition, New Delhi, Oxford University Press.
2. Bhushan Barat, Kumar Vinit. (2006) *Environmental Impact Assessment*, New Delhi, Shree Publishers & Distributors.
3. Birnie, Boye,& Redgwell. (2009) *International Law and the Environment*, New York,Oxford University Press Inc.
4. Chowdhary Nupur. (2016) Sustainable Development as Environmental Justice Exploring Judicial Discourse in India, Vol LI NOS 26 & 27, *Economic & Political Weekly* pp 84-92.
5. Chowdhary Nupur. (2014) *Environmental Impact Assessment in India: Reviewing Two Decades of Jurisprudence 5 IUCNAEL E Journal*, pp 28-32. http://www.iucnael.org/en/86-journal/issue/491-issue-20142.
6. Cullet Phillipe. (2005) *Intellectual Property Protection and Sustainable Development* New Delhi, LexisNexis Butterworths.
7. Kothari Ashish, (2002) "Sustainable Development" *The Hindu Survey of the Environment*, P. 27.
8. Kumar Swatanter. Access to Environmental Justice in India and Indian Constitution' *NGT International Journal of Environment Vol.1 (2014)*.
9. Leelakrishnan P. (1999) *Environmental Law in India* New Delhi Butterworths India.

10. Ministry of Environment and Forest, (2006) National Environment Policy, New Delhi: Government of India.
http://www.moef.gov.in/sites/default/files/introduction-nep2006e.pdf
11. Nomani, Zafar Mahfooz. (2010), *Environment Impact Assessment Laws*, New Delhi, Satyam Law International.
12. Nomani, Zafar Mahfooz. (2004). *Natural Resources Law and Policy*, New Delhi, Uppal Publication.

CASES CITED

1. Narmada Bachao Andolan v Union of India (2000): SCC, SC,10, p.664
2. *ND Jayal and Another v Union of India and Others* (2004): SCC,Sc, 9, p 362.
3. *Utkarsh Mandal v Union of India* (2009) Judgment of Delhi High Court on Novenber 26,,2009.
4. *Lafarge Umiam Minig Private Limited v Union of India & Ors* (2011): SCC, SC, 7, p 338.
5. *Gram Panchayat Navlakh Umbre v Union of India and Ors,* (2012): Judgment of Bombay High Court on June 28,2012.
6. *Deepak Kumar v State of Haryana and Ors,* (2012): Judgment of Supreme Court on February 27,2012.
7. *Sterlite Industries (India) Ltd. v Union of India* (2013): AIR 2013 SC p.3231.
8. *G Sundarrajan v Union of India and Ors* (2013): SCC, Sc, 6, p 620.
9. *T.N Godavarman v Union of India* (2014): Order of the Supreme Court on January 6,2014.
10. *Talaulicar & Sons P. Ltd. v The Union of India* (UOI) and Ors.(2016): MANU/GT/0761/2016.
11. *Electrotherm (India) Ltd. v Patel Vipulkumar Ramjibhai and Ors* (2016): MANU/SC/0850/2016, AIR 2016SCp.3563.
12. *Common Cause and Ors. v Union of India and Ors* (2107): MANU/SC/0930/2017.
13. *Adivasi Majdoor kisan Ekta Sangathan and Another v. Ministry of Environment and Forest and Others* (2012): Appeal No. 3/2011, Judgment passed on April 20, National Green Tribunal, Principal Bench, New Delhi.
14. *Antarsingh Patel and Ors v Union of India* (2012): Appeal No. 26/2012, Judgment passed on 9 August, National Green Tribunal, Principal Bench, New Delhi.
15. *Gau Raxa Hitraxak Manch and Gaucher Paryavaran Bachav Trust, Rajula v Union of India and Others* (2013): Appeal No. 47/2012, Judgment passed on 22 August, National Green Tribunal, Principal Bench, New Delhi.

16. *Samata and Forum of Sustainable Development v Union of India & Ors* (2013): Appeal No. 9/2011, Judgment passed on 20 April, National Green Tribunal, Southern Zone Bench, Chennai.
17. *M P Patil v Union of India* (2014) Appeal No. 12/2012, Judgment passed on 13 March, National Green Tribunal, Principal Bench, New Delhi.
18. *T Muruganandam and Ors v Union of India and Ors* (2014): Appeal No. 50/2012, Judgment Delivered on 10 November, National Green Tribunal, Principal Bench, New Delhi.
19. *S.P. Muthuraman and Ors. v Union of India and Ors*. (2015): Appeal No. 37/2015, Judgment Delivered on 07 July, National Green Tribunal, Principal Bench, New Delhi.
20. *Vikrant Tongad v Noida Metro Rail Corporation and Ors,* (2016): Original Application No. 478/2015, Judgment Delivered on 31 May, National Green Tribunal, Principal Bench, New Delhi
21. *Santhiyagu v Union of India and Ors.* (2017): Appeal No. 66/2016, Judgment passed on 05 May, National Green Tribunal, Southern Zone Bench, Chennai.

Chapter 13

Environmental Impact Analysis: A Socio-Legal Study of Kol Dam in Himachal Pradesh

Dr. Kailash Thakur & Dr. Harish Thakur

Abstract: *Constructions of dams have been viewed significant and inevitable to solve various social and economic problems of the country. Since Independence, the Central and State Governments' of India are undertaking large dams based on these assumptions. Governments are still in the process of damming most of the rivers in India despite of the fact that already constructed dams' have failed miserably to deliver their optimal. The Governments' have been unsuccessful to take into consideration their precise environmental, social and legal dimensions. In most of the cases of dams induced displacement, the affected people have not been rehabilitated and resettled till date. The adverse environmental consequences of large dams can also not be denied. Having this background in mind, the objective of the present paper is to make analyses of the social, legal and environmental impacts of the kol dam in State of Himachal Pradesh. The authors also tender some suggestions how adverse consequences of large dams can be minimized and addressed to. The present study is empirical in nature and is based on primary data gathered from affected areas of the Kol dam.*

Keywords: *Dams, Displacement, Social, Legal, Environment, Impacts.*

13.1 INTRODUCTION

Since Independence, Central and State governments' are undertaking large dams in India. Governments' consider building of large dams inevitable to resolve the issues of food insecurity, irrigation, poverty, unemployment and flood etc, the problems with which India is struggling. But individual studies have been indicative that the dams have failed to achieve these objectives. These challenges still continue undefended. Rather construction of large dams has generated environmental, social and legal impacts. Thousands of people are forced to give up their homes, traditional occupations and common property resources in the name of national development. The dams have disconnect, people from their social networks, past culture, traditions, customs and religions, etc.(Ayesha Pervez,2008). Individual study has pointed out several impoverishments risks of forced displacement (Michael M. Cernea, 1996). Not only people have been socially affected but environment has also been damaged to great extent. Dams have their own downstream and upstream disadvantages (Balgovind Baboo, 1991). Besides, dams affected wild life, flora and fauna and natural water resources of the areas where built. The construction of dams by governments without carrying Environmental Impact Assessment (in short EIA) and Social Impact Assessment (SIA) studies further aggravates the problems.

The erecting of Kol dam hydroelectric power project on river Sutluj in Himachal Pradesh has also raised several environmental, social and legal issues. The dam has been built by government of Himachal Pradesh in collaboration with National Thermal Power Corporation (NTPC). For construction of this dam large area of forest has been cleared off. Also, as per official information total 1100 families have been displaced due to dam. Like governments of different states, Himachal Pradesh government had also undertaken Kol dam based on several assumptions as citied above. But the field survey of the areas affected painted entirely different picture. Not only the environment of the areas got affected but various socio-legal impacts of the dam also become evident.

13.2 OBJECTIVES & SAMPLE OF STUDY

The Universe of the present study is the entire gamut of Project Affected Families (Oustees) of Kol dam.

   i. To evaluate socio-legal impacts of Kol dam hydro power project in light of data collected.

ii. To suggest measures how environmental, social and legal repercussions of the construction of dam can be minimized and addressed to.

In order to assess the socio-legal impacts of the Kol dam, total 330 project affected families have been interviewed. Quasi participant observation method and interview method of research both have been used to record responses of the oustees for the completion of present study.

13.3 SOCIAL IMPACTS OF KOL DAM
13.3.1 Disintegration of Joint Family System
The first significant social impact of the Kol dam induced displacement has been on family set up of the affected villages. In affected villages, prior to displacement most families were joint which reduced into nuclear families in post displacement period. This conversion and shift was not by choice. The respondents interviewed stated that they wanted to live in joint family system because of its several advantages. **Fig.1** reveals that 236 out of 330 (71.5 percent) families were joint and the remaining 94 out of 330(28.5 percent) families were nuclear or separate in pre-displacement period. However, in post displacement period only 55 out of 330(16.7 percent) families are now joint families, while 275 out of 330(83.3 percent) followed the separate or nuclear family system. Thus, Joint family system has considerably reduced to nuclear family system in post displacement period.

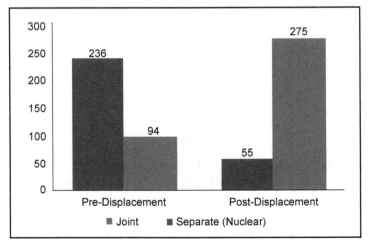

Fig. 13.1:

13.3.2 IMPACTS ON EARNING SOURCES:

In pre-displacement period, the agriculture was main source of income which majority people lost and could not continue in post displacement period. The people affected in rural areas had expertness in farming and agriculture and they expressed their willing to continue this occupation in post displacement period, but could not continue because of loss of entire land. Even the NTPC and Government of H.P have not provided land based rehabilitation. This has aggravated the problems of those families who were entirely dependent on agriculture. **Fig-2** reveals that for 292 out of 330(88.5 percent) families, agriculture was the main source of income which reduced to 57(17.3 percent) in post displacement period. Oustees after losing traditional sources of income started to do casual labor work in post displacement period. It goes up to 65(19.7 percent) from 17(5.1 percent) in post displacement period.181 out of 330(54.9 percent) displaced persons had virtually no source of income. Thus, the data clearly indicates that in post displacement period the sources of family income drastically affected and reduced to a greater extent. It has adversely affected the livelihood of the families and amounts violation of a right to earn livelihood encapsulated in Art.21 and right to carry on traditional occupation guaranteed under Art. 19(1) (g) of Indian Constitution.

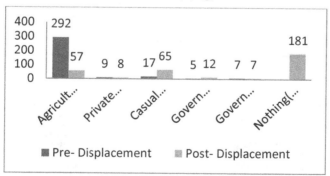

Fig.13.2:

13.3.3 Women's Dependency Increased:
Government's assumption and plea that undertaking of large dams will help in empowering women in terms of employment, education and health proved wrong when women affected in Kol dam entirely become dependent on their male spouse in post displacement period.

They lost their hold on common property resources and traditional occupation. The schools of the areas and health facilities affected to larger extent. The women had no say in fixing the rate of their assets' and using compensation amount for their betterment. Women lost their earning sources completely. **Fig.-3** shows that 178 out of 330(53.9 percent) women oustees were earning their livelihood from the agriculture but in post displacement period only 39 out of 330 (11.9 percent) women oustees could continue this traditional occupation. Now they are forced to earn their livelihood by doing casual work in various households, factories, shops and some in kol dam project.

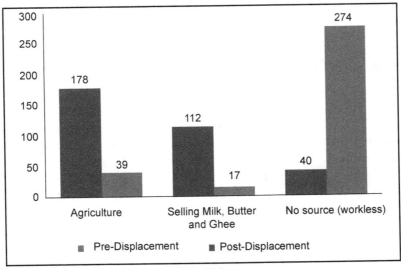

Fig. 13.3:

They are totally dependents on their male spouse to meet their day to day expenses. Majority women interviewed stated that they would like to continue their traditional occupation in post displacement period.112 out of 330(33.9 percent) women possessed livestock and were earning their livelihood from animal husbandry before displacement. But the number of women earning such income had reduced up to 17 out of 330(5.1 percent) in post-displacement period. Similarly 274 out of 330(83.0 percent) women in the paucity of the source of income or work they have had before displacement are now house wives in post displacement period without any sources of earning income. Loss of agriculture land and reduction of livestock obviously are the reasons for rendering them as workless in post displacement period.

13.3.4 Impacts on Surrounding Environment:

As per official information, large forest area was cleared off for the construction of dam. The deforestation has caused problem of soil erosion and several other environmental problems in the affected areas. The afforestation compensatory scheme of the NTPC and Government of Himachal Pradesh has not been implemented seriously. Besides, natural water resources were dried up due to heavy blasting in the area. The dust particles settled down on leaves resulting in blockage of photo synthesis process. The substantial number of respondents stated that Kol dam adversely affected the surrounding flora, fauna and people. The wild animals were largely affected.

13.4 LEGAL IMPACTS OF KOL DAM

13.4.1 Constitutional Issues

Several constitutional and legal repercussions of Kol dam come to light during displacement and their subsequent rehabilitation and resettlement. People ousted were not consulted prior to their displacement. The police personnel used *lath charge* against the people who were holding peaceful demonstrations against construction of dams and acquisition of their land at rate less than prevalent market rate. During the entire process of land acquisition and allotment of plots, respondents faced various challenges and most of the times feel humiliated. All these actions of the governments and their instrumentalities raised constitutional issues of right to live with human dignity, right to be informed, right to take consensual decision, right to life and personal liberty etc.

Fig- 4 reveals that 221 out of 330(67.0 percent) respondents opposed the project, while only 89 out of 330(27.0 percent) were in favour of the project and remaining 20 out of 330 (6.1 percent) did not reply the question. The respondents were further asked to inquire in what way they opposed the project. Fig further reveals that an overwhelming number 107 out of 221(48.4 percent) held peaceful demonstrations while 74 out of 221(33.5 percent) adopted the methods of protest meetings. The respondents stated that their protest movements were brutally suppressed by govt. **Fig.5** indicates that 91 out of 221(41.2 percent) stated that police force has been used to suppress protest movements and to take over the possession of land and houses. While 7 out of 221(3.1 percent) respondents uttered that false criminal cases were filed against them by NTPC officials. Another 58 out of 221(26.3 percent) respondents reported that project authority gave them the allurement of jobs in the project and 47 out of 221 (21.3 percent) respondents were arrested and *lathi* charged while protesting against the project. Remaining 18 out of 221(8.1 percent) respondents affirmed that they were threatened by the project authority that their compensation amount would be forfeited if they would oppose the project.

The forcible uprootment of people from their land and house raises serious question of rule of law. The methods adopted by Govt. of H.P to oust oustees are against the Constitution of India and various international human rights laws, principles and conventions governing rights of oustees.

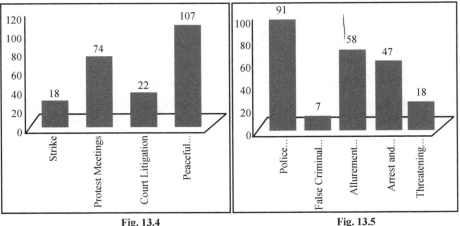

Fig. 13.4   Fig. 13.5

13.4.2 Issue of Gender Discrimination
An empirical survey revealed that women affected due to dam were not consulted before acquisition of land. Majority women had no idea how much compensation their male partners had been paid. The payment cheques were issued in the name of male member in the family. The female oustees had no say. It increased their dependency on their male spouse and made their social and economic status secondary. In matter of payment of compensation and even during distribution of resettlement and rehabilitation packages and grants, the women were not consulted.

Table-1 reveals that 240 out of 330 respondents were not informed regarding R & R Scheme of the govt. of H.P which indicates that project authority did not inform the oustees on important matters. It is denial of a right to be informed which has been recognized a significant human right. Again only 34 out of 330(10.3 percent) women were aware about the rehabilitation and compensation packages and an overwhelming number 285 out of 330(86.4 percent) were not aware about the available R&R packages. It depicts that the women oustees were treated secondary and they were discriminated which amounts violation of a right to equality.

4.3 Forcible Resettlement of Oustees: Freedom to choose residence is one of the right available to citizens of India under Indian constitution. This right is available to oustee also. But project affected people of Kol dam were forcibly settled at plots which were inadequate and incomplete. The allotted plots were rejected by many oustees due to their size and inadequate development. Most of the families rejected plots allotted because of the lack of basic amenities at resettlement colonies. The water scarcity, inadequate health facilities and dysfunctional and inadequate sewerage and sanitation facilities at resettlement colonies were the major problems.

Table-1 specifies that 71.8% oustees were displaced arbitrarily from their land and house. The government did not take displaced into confidence before serving notices to give up possession of their land. The procedure given to oust people under the Land Acquisition Act is unconstitutional and government failed to note it. The project authority applied arbitrary procedure to evict the villagers from their land and house and thus violated Constitutional right to property guaranteed under Article 300-A. The majority of the ousted families were not told about the purpose of acquiring their land. Similarly this move of the government also amounts violation of the right to be protected against arbitrary displacement, right to be informed and participation of displaced families in decision making. All these human rights are recognized in UN Guiding Principles on Internal Displacement, 1998, ILO Convention of 1969 for Indigenous Peoples, World Bank Policy of 2001 and Asian Development Bank Policy on Involuntary Resettlement, 2006.

Table shows that 79.7% ousted families were not consulted prior to their relocation. The families had forcibly resettled on relocation site allotted. In majority cases resettlement sites had not been selected in consultation with displaced persons. It amounts the violation of right to choose one's own residence and right of self-determination guaranteed in our Constitution and several human rights laws.

Table 13.1: Awareness among Oustees about undertaking of Kol Dam Project

Indicators	Yes	No	No Response	Total
Consulted for displacement	74 (22.4%)	237 (71.8%)	19 (5.8%)	330 (100.0%)
Consulted about relocation sites	59 (17.9%)	263 (79.7%)	8 (2.4%)	330 (100.0%)
Informed about R&R Scheme of the govt. of H.P	84 (25.5%)	240 (72.7%)	6 (1.8 %)	330 (100.0%)
Awareness of women about rehabilitation and compensation packages	34 (10.3%)	285 (86.4%)	11 (3.3%)	330 (100.0%)

13.4.4 Issue of Inadequate Compensation

Under the Land Acquisition Act of 1894, there was no provision for land based rehabilitation. It only had a provision for compensation. The compensation based rehabilitation has several repercussions. Major among them are, there is chances of misappropriation of compensation amount especially by villagers, compensation is not fairly calculated and women remains unaware about compensation amount given. Furthermore, majority respondents were not happy and satisfied with the way compensation was paid and calculated. Only 13 out of 330(3.9 percent) respondents found the compensation adequate to rebuild previous standards of life. An overwhelming number 272 out of 330(82.4 percent) respondents held that it was not adequate and it could have been more. Another 45 out of 330 (13.6 percent) did not respond the question.

Table 2 shows that only 19 out of 330 (5.8 percent) respondents were satisfied with the way of determination of compensation while majority respondents interviewed i.e. 246 out of 330(74.5 percent) complained that they are not satisfied with the way the compensation amount was determined. Another 65 out of 330 (19.7 percent) were unaware about the fixation of compensation amount and did not respond the query. Thus majority i.e. 74.5 % respondents were not satisfied with the way the compensation for land; house and standing trees was calculated by government. It was paid in lump sum and not in accordance to the prevalent market rates. Moreover, compensation was disbursed very late. All these facts revealed that in matters of calculating compensation the government adopted different parameters and did not calculate it fairly. It amounts violation of a human right to get fair and just compensation for lost assets.

Table 13.2:

Sr. No.	Satisfaction level about determination of Compensation	Frequency	Percentage
1.	Yes	19	5.8
2.	No	246	74.5
3.	Don't Know	65	19.7
	Total	330	100.0

13.4.5 Arbitrary disbursement of compensation Amount

It has been discussed above that women's rights have been rarely respected. They suffer in terms of money, status and decision making. Data in **Table 13.3** reveals that 321 out of 330(97.2 percent) oustees the cheques of compensation amount were handed over to the head male member of the family and only in 9 out of 330(2.8 percent) cases cheques were allotted to the female head of the family. It is amply clear that women dam oustees were discriminated in matter of receiving compensation. It raises the questions of gender discrimination and amounts violation of a right against discrimination and a right to equality.

Table 13.3:

Sr. No	Recipients of Cheques	Frequency	Percentage
1.	Male head of the family	321	97.2
2.	Female head of the family	9	2.8
3.	Both(Male/Female)	0	0
	Total	330	100.0

13.4.6 Issues Relating to Religion

In survey area, the people were strong religious customs and practices. They were used to worship their deities at various occasions. Displacement severely affected religious sentiments of the people. The temples of ousted families were demolished deliberately and without informing them. This deliberate and malicious act of executing authority was intended. It outraged religious feelings of oustees resulting into violation of Indian constitution which guarantees to all religious rights. This arbitrary action of the NTPC and Govt. of H.P amounts violation of cultural and religious rights as incorporated in articles 25-28 of Indian constitution.

Fig.13.6 indicates that approximately 91.9 percent of oustees were not informed well in advance to demolish their common temple like temple of *Isht or Gram dev* and *kulja devi*. Only, 5.9 percent of project affected families were informed.

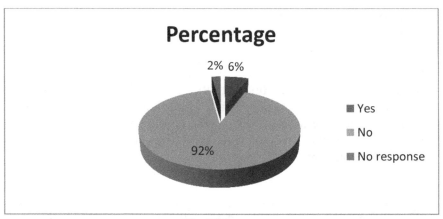

Fig.13.6:

13.4.7 Policy on Acquisition & Allotment of Houses

Sudden and uninformed displacement of people from their homes caused great mental agony to the displaced. The project authority did not follow proper procedure to oust people. It is needless to mention here that Land Acquisition act of 1894 did not satisfy the requirement of due process. As a result of this, the respondents forced to stay in temporary houses where none of their rights were safe. The women and young girls had the issue of privacy and they were forced to urinate in open or temporary made houses of sheets. During initial years of displacement, the residents were at risk. There was no proper planning and provision for resettlement of people before their actual physical displacement. It raised serious legal issues.

Table 13.4 indicates that majority of the respondents i.e. 199 out of 330(60.3 percent) initially resided in temporary houses made of tin sheets. Another 93 out of 330(28.2 percent) stayed in rented house at different places. Similarly 25 out of 330 (7.6 percent) persons stayed in their own houses at any other place and only 13 out of 330(3.9 percent) respondents informed that they stayed with their relatives at different places after displacement. The reasons of oustees for staying at different places was that resettlement colonies were rejected by majority of the dam oustees as pointed out in **Table 13.5.**Table indicates that an overwhelming number 245 out of 330 (74.2 percent) dam oustees rejected resettlement colonies. These 245 displaced relocated and resettled on their own without assistance of the government and project authority. They did not opt to reside at resettlement colonies because resettlement colonies built for them lack basic infrastructural facilities.

Thus, majority of the oustees and their families were forced to reside in tin made or make shifts houses initially for one year. The oustees were forced to live in unhygienic conditions and face various problems regarding their privacy, space and toilet in these tin made houses. It amounts denial of right to privacy and right to live in healthy environment.

Table 13.4:

Sr. No.	Stay during Initial Period of 1 year of displacement	Frequency	Percentage
1.	In temporary make shift homes	199	60.3
2.	In rented house	93	28.2
3.	Own house at any other place	25	7.6
4.	With relatives	13	3.9
	Total	330	100.0

Table 13.5:

Sr. No.	Whether Residing at Resettlement Colonies ?	Frequency	Percentage
1.	Yes	85	25.8
2.	No	245	74.2
	Total	330	100.0

13.4.8 Dam failed to provide Employment

Employment based rehabilitation can help to minimize repercussions of displacement. But it has been seen that once the people are displaced, government ignore employment of oustees in project. It raises several legal issues. The empirical survey reveals that 261 out of 330 (79.1 percent) of the displaced families did not get any employment in the project. Only 69 out of 330(20.9 percent) respondents got employment of temporary nature in the project.

Study reveals that (i.e. 98.6) family member got temporary or casual work in the project which entirely gave an opportunity to contractor to exploit them. The persons whom employment of temporary nature had been given could continue it. 54 out of 69(78.2 percent) are not working in the project and only 15 out of 69(21.8 percent) displaced persons are still working in the project. Those oustees who did not opt to continue their jobs in the Kol dam project were further asked the reason of leaving jobs.

Empirical survey further reveals that 29 out of 54 (53.7 percent) reported that no work was available in the projects. Similarly, 14 out of 54(25.9 percent) respondents reported that they left work owing to the reasons of low wages and exploitation by private contractors. Again 8 out of 54(14.8 percent) respondents reported that their services were terminated by the contractors without stating any reasons. Due to unavailability of work in post displacement period oustees were readily offering their services. The contractors take advantages of this situation, and employed cheap labour in the project at different work. But it is unfortunate here to note that these contractors were not paying full wages to the oustees employed by them. They were exploiting less educated oustees by not giving the salary on time. These arbitrary actions of the authorities' amounts violation of a right against exploitation which is a significant human right enforceable even against private individuals.

13.4.9 Other Legal Impacts of Kol Dam

Table 13.6 depicts that 15 out of 330 (4.5 percent) of the respondents stated that Kol dam project disturbs the education of girl child of the families. Another 55 out of 330 (16.6 percent) respondents indicated loss of livelihood or work opportunities as impact of displacement; 32 out of 330 (9.7 percent) held loss of access to CPR's as major impact of displacement. It is worthy to mention here that the rural populations in India have the right to use, plan and manage common property resources such as natural water, rivers, fodder, pasture, grazing land, fuel wood, forest produce, medicinal plants and herbs etc. Some of the respondents assigned other impacts of displacement on them and their families such as 5.4 percent respondents held the view that women of the family lost the source of earning livelihood and 2.1 percent thought that breakdown of social bonds and ties is the major impact of displacement. Similarly 29 out of 330 (8.8 percent) respondents indicated that displacement is responsible for the deterioration of health. Another 70 out of 330 (21.2 percent) respondents stated that loss of good quality agricultural land has been the major impact of displacement on them and their families. A substantial number of the respondents i.e. 104 out of 330 (31.6 percent) stated that all the above mentioned impacts are noticed after displacement on them and their families.

Thus major impacts of the Kol dam have been on health, education, livelihood, and common property resources of the villagers. It amounts violation of right to health, right to education and right to access to common property rights which are human rights recognized at national and international level.

Table 13.6:

Sr. No	Effect of Kol Dam Project	Frequency	Percentage
1.	Disruption of Girls' Education	15	4.5
2.	Loss of work	55	16.6
3.	Failure to access to CPR's	32	9.7
4.	Loss of Women's' earnings	18	5.4

Contd...

5.	Weakening of social bonds	7	2.1
6.	Loss of health	29	8.8
7.	Loss of Agricultural land	70	21.2
8.	All the above	104	31.6
	Total	330	100.0

13.5 CONCLUSIONS AND SUGGESTIONS

The foregoing discussion reveals that there have been several social and legal impacts of construction of Kol dam. It resulted in unprecedented displacement of families and put adverse impacts on surrounding environment. The project affected people of the areas lost their traditional occupation, livelihood and social bonding. The people have been denied basic amenities like water, electricity and sanitation at resettlement colonies. The wild animals of the areas have disappeared. The flora and fauna has largely affected. The water resources, streams and rivers of the area have become polluted and dried. The Government and NTPC did not conduct the environmental and social impact assessment study prior taking up of the project. This lackadaisical approach on their part has further aggravated the problems of oustees.

Empirical survey also revealed that women oustees and their children have been deprived of basic rights like right to education, right to family, right to feed, right to health and right to enjoy recreational facilities etc. This vulnerable section faces sever discrimination in matter of payment of compensation. The Constitutional and ordinary rights of the people have least respected which raises serious concern in India where governments' makes tall claims to protect people's rights. Several legal issues left unaddressed during dealing with oustees and their subsequent rehabilitation. Keeping in view the adverse consequences of Kol dam, the present study submit following suggestions to minimize environmental, social and legal impacts of the dam.

    i.   Studies have been indicative that large dams have more devastating impacts as compared with small dams. Therefore, instead of undertaking large dams, it is suggested to undertake small dams to minimize environmental, social and legal impacts of displacement.

ii. It is also suggested that before planning to construct a dam, the people who are going to be displaced and affected need to be consulted. Prior consultation with people will generate a confidence and trust among their mind towards Governments and project executing authorities. The people would feel privileged that they are contributing in development of the country. This step will also help in minimizing pain of uprootment and prepare people mentally to shift to new places.

iii. The project authorities and Govt. before displacing people should properly develop resettlement colonies. All steps should be taken well in advance to provide basic facilities like health, education, sanitation, electricity and water at resettlement colonies to minimize discomforts and pain of people.

iv. In matter of determination and awarding compensation, the women oustees should also be consulted. They should not be discriminated. Adequate opportunities should be given to them to express their view point on undertaking of projects and their adverse impacts on their livelihood and other income earning sources. They should made participant in decisions making, allotment of compensation, house, financial grants and jobs in the project. Women, participation in the entire process of resettlement planning and implementation should be ensured to protect their right to equality.

v. The land and employment based rehabilitation should be promoted.

vi. The education of Children of displaced families especially education of a girl child gets materially disturbed during the course of displacement. It is suggested that the expenses of the school going children should be borne by the government until the family of displaced children is properly rehabilitated.

vii. The task to distribute all types of assistance should be in the supervision of properly constituted body consisting of government officials, members of NGO's and

	representatives of displaced families to prevent discrimination in receiving these benefits.
viii.	The waste and degraded land should be acquired at the place of good agricultural land for the construction of dams.
ix.	The project authority and government should rehabilitate the displaced communities in groups at resettlement colonies situated near to each other and connected with roads. This arrangement would help to build lost social networks among the displaced families in post displacement period.

REFERENCES

1. Ayesha Pervez,(2008), "Gendering the National Rehabilitation Policy of India," *Social Change*, p.52.
2. Michael M. Cernea,(1996) "Public Policy Responses to Development-Induced Population Displacements," *Economic and Political Weekly*, Vol. 15. p.1518.
3. Balgovind Baboo,(1991) "Development and Displacement: The Case of Large Dams in India," *Man and Development*, p.35.

Chapter 14

Sustainable Mining and Closure Policy Regulations and Practice: A Case Study of Coal Mining in Meghalaya

Ali Reja Osmani

A technological society has two choices. First, it can wait until catastrophic failures expose systemic deficiencies, distortion and self-deceptions Second, a culture can provide social checks and balances to correct for systemic distortion prior to catastrophic failures.

Mahatma Gandhi

Abstract: *The objective of this paper is to highlight India's policy framework regarding mining sector, sustainable mining and mine closure in particular. The present paper is important because sustainable mining and mine closure being a serious problem around the globe, India introducing stringent regulation to handle the address environmental concerns and to analyse these norms in the context of mining in Meghalaya. The early mining operations has adversely affected the environment by way of contaminated surface water and groundwater, ugly and unstable tailings with acid rock drainage, and dangerous open adits, have conferred the mining industry with both fame and defame. Paper discusses the concept of mine closure and its development over the years and also the contribution of 'Social License', 'Financial Assurance', standards set by 'International Financial Bodies', etc. Paper also tries to highlight the existing mining scenario in the state of Meghalaya with a brief overview of the mining sector in the region. In addition this* *part shall also discuss the Economic and Livelihood concerns raised due to the April 17, 2014 ban on "rat-hole' coal mining in Meghalaya by the National Green Tribunal (NGT) through an interim order. The paper also tries give an overview of the entire legal framework for mining in the country by discussing a brief history of the mining sector, amendment made in respect of mine closure and the integrated approach to adopt the sustainable development framework. It is difficult to arrive at any conclusion regarding the effectiveness of policies. Mining processes are time taking processes, mostly continues for many decades and only at the end of the mine life cycle an evaluation could be possible. The mining sector is in the rise in India and the policies of prospecting mine closure plan and post closure along with other environmental jurisprudence will help the mining sector to be sustainable sector. Keeping in view the approach for sustainability in mining sector there is more stringent provisions likely to come under the preview of legal framework in the years to come.*

Keywords: *Sustainable Mining, Mine Closure, Coal Mining, Meghalaya*

14.1 INTRODUCTION

By its very nature, mining cannot be regarded as a form of sustainable development. But contrary to this fact, if sustainable development is defined as a balance between social, economic and environmental considerations, then a mining project that is developed, operated and closed in an environmentally and socially acceptable manner could be considered as promoting sustainable development (Sessoon, 2000).The early mining operations has adversely affected the environment by way of contaminated surface water and groundwater, ugly and unstable tailings with acid rock drainage, and dangerous open adits, have conferred the mining industry with both fame and defame (Garcia, 2008, p. 2).Historically in mining practices, two phases in the typical mining project operational life cycle has been recognized such as exploration and exploitation/production and therefore the focus was always on the immediate requirement of those stages. But in present day, the mining projects progress through a life cycle, which can be classified as such, (i) Exploration (ii) Pre-feasibility assessment (iii)

Construction (iv) Production and (v) Closure (vi) Post-closure (Morrey, 1999).

Ever since the term ‚Sustainable Development' has been coined by ‚Brundtland Commission' in its report ‚Our Common Future' in 1987, it has been on the forefront of the global agenda. With the recent adoption of the post-2015 Sustainable Development Goals (SDG) at the UN General Assembly in September 2015 to carry further the targets of Millennium Development Goals, which was adopted in 2000 by world leaders. Some targets set by SDGs are very much relevant from the perspective of this paper. The mining industry can impact both positively and negatively across the SDGs. It is necessary to mention here that in recent times, the industry has made significant advances in improving how mining sector manage their environmental and social impacts, protect the health of their workers, achieve energy efficiencies, respect and support human rights, provide opportunities for decent employment conditions and foster economic development. However, it is also well settled that mining has contributed to many of the challenges that the SDGs are trying to address now, such as environmental degradation, carbon emission, displacement of populations, worsening economic and social inequality, armed conflicts, gender-based violence, tax evasion and corruption, and increased risk for many health problems etc. Given the existing negative and positive impacts of mining combined with the industry's capability to mobilize human, physical, technological and financial resources, the mining sector around globe potentially has direct or indirect relations with all the 17 of the SDGs where the sector can contribute and ideally can position itself as a leader in advancing the SDGs. However most relevant among them are *SDG-1: End Poverty* — Mining generates significant revenues through taxes, royalties and dividends for governments to invest in economic and social development;*SDG-6: Clean Water and Sanitation and SDG-15* — Life on Land: Mine development requires access to land and water presenting significant and broad landscape impacts that must be responsibly managed;*SDG-7: Energy Access and Sustainability and SDG-13: Climate Action*— Mining activities are energy and emissions intensive in both the production and downstream uses of its products;*SDG-8:Decent Work and Economic Growth* — Mining can change the lives of local communities, offering opportunities for jobs and training, and also contributes to economic and social inequities if not appropriately managed; *SDG-9: Infrastructure, Innovation, and Industrialization* — Mining can help drive economic development and diversification through direct and indirect economic benefits, the development of new technologies and by spurring the construction of new infrastructure for transport, communications, water and energy; and*SDG-16: Peace, Justice and Strong Institutions* — Mining can contribute to peaceful societies by avoiding and remedying company-community conflict, respecting human rights and the rights of indigenous peoples, and by supporting the representative decision-making of citizens and communities in extractives development.

The objective of this paper is to highlight India's policy framework regarding mining sector, sustainable mining and mine closure in particular. The present paper is important because sustainable mining and mine closure being a serious problem around the globe, India introducing stringent regulation to handle the address environmental concerns and to analyse these norms in the context of mining in Meghalaya. In this paper, after the introductory part, the Section-2 discusses the concept of mine closure and its development over the years and also the contribution of ‗Social License', ‗Financial Assurance', standards set by ‗International Financial Bodies', etc. The Section-3 shall try to highlight the existing mining scenario in the state of Meghalaya with a brief overview of the mining sector in the region. In addition this part shall also discuss the Economic and Livelihood concerns raised due to the April 17, 2014 ban on ―at-hole' coal mining in Meghalaya by the National Green Tribunal (NGT) through an interim order. The Section-4 contextualize the entire legal framework for mining in the country by discussing a brief history of the mining sector, amendment made in respect of mine closure and the integrated approach to adopt the sustainable development framework. Finally Section-5 contains the concluding remarks. The methodology adopted is mostly employs legal and policy analysis of the Indian mining sector. Primary sources are reviewed to identify the areas of analysis and secondary sources are consulted to substantiate the author's own opinion. The study based on various sources, most of the sources are Government of India reports, working papers and national legislations and with few International best practice models. In this study, some sources belong to International and National organization and the remaining sources are books, journals and articles are also relied on.

14.2 CONCEPT OF MINE CLOSURE

If we look back couple of decades, mine closure for a mining company primarily meant, removal of equipment for salvage or sale, collection of accounts receivable, laying off its workforce, and extinguishing its legal title. With growing environmental concerns and awareness, attention moved towards reclamation, initially with regard to pollution containment, health and safety but subsequently focusing on ecological system restoration. Today, social and economic issues are being considered, particularly in places where mines have considerable economic influence to the local and/or national economy (Otto, 2009, p. 252). Closure planning must integrate and incorporate all aspects of sustainable development. Environmental, economic and social issues must be addressed in a just manner in the closure plan and incorporation of such integrated plan must continue throughout the mine life (Mining for the Future, 2002, p. B-8).

Till 1960s very little initiative were taken for activities beyond operational phase and it's only in 1970s, the need for decommissioning and surface reclamation was recognized (Hodge & Killam, 2003, p.4). The term reclamation and closure are not synonymous. The word reclamation refers to the process whereby a mine's landform and ecology are altered to attain planned outcome. Whereas the word closure refers to actions that includes physical shutdown of mine site and also includes plenty of other activities, such as final reclamation, equipment removal, community disengagement, employee severance, debt settlement etc., when company decides to stop its operation (Otto, 2009). Therefore, there are two aspects of mine closure plan'---- the objective of mine closure is to leave a mine site in a condition which is safe and stable, and to put an end to further environmental degradation so that the mining tenement can be renounced for alternative land use and the objective of land reclamation is to restore the land affected by mining to enable it for another economic use in future (Hoskin, 2005). The planning for closure would require that the time frame for Environmental Impact Assessment (EIA) be extended not only to time of closure but even to the post closure phase (Warhurst & Noronha, 1999, p. 8).

The major objectives of mine closure are as follows (Trivedi, R., et.al., 2011, p. 31):
   a) To safeguard health and safety of the public,
   b) Land after closure shall not be affected and shall ensure sustainability in long run,
   c) Environmental resources in the area shall not be degraded in any form,
   d) Minimize adverse socio-economic impacts,
   e) Develop potential for communities' future prospect in respect of economic and social life.

The selection of closure options are based upon the effectiveness, risk of failure, long term stability and cost factors. Existing trends in closure planning and design involve technical review and assessment of risk and cost benefit, both in respect of engineering and environmental terms. Such review and assessment assists in selecting the most appropriate design alternative (Morrey, 1999, p. 16). A good mine closure plan should encompass the following priorities in consideration such as (World Bank & International Finance Corporation, 2002, p. 7):
   a) a clear idea about time horizon and costs,
   b) identifies the expected final landform and surface rehabilitation, including removal of plant and equipment and stabilization and detoxification of dumps and impoundments,
   c) make risk assessment to set priorities for preparatory work,
   d) make cost benefit analysis of different options at the planning stage,
   e) gives a management plan for the implementation of closure, and
   f) also include proposals for post closure monitoring.

Therefore the final closure should include multidimensional activities at this stage to become a comprehensive mine closure plan such as, (i) removal of infrastructure, (ii) the implementation of public safety measures, (iii) rehabilitation, (iv) continuing maintenance of site structure and monitoring of environmental issues, (v) the operational activities at the site facilities are required to mitigate or avoid long term environmental degradation, (vi) company's involvement in sustainable community economic and social measures (Mining for the Future, 2002, p. B-3).The entire concept of mine closure as a whole not very old, the idea of ‚design for post closure' is a recent addition to the mine closure debate. In 1983, for the first time ‚design for closure' was conceived as part of the mining project in Cinola Gold Mine located on the Queen Charlotte Islands off the coast of northern British Columbia (Hodge, and Killam, n.d., p. 14). So the concept of mine closure has developed over the years from the initial exploration and exploitation stages of the mining industry to the concept of mine closure and further the concept of post closure. Social license, financial assurance and standards prescribed by international financial bodies also contributed in the continuous evolution of the concept in addition to the efforts made by the mining industries itself.

14.2.1 Social License

The historic mining operations that were abandoned without closure have affected the environment negatively and influenced the ability of mining firms to obtain a social license. Companies gradually started to learn that they must plan for closure as part of the initial phase of exploration for two reasons, (i) because the communities demand that a mine should obtain a social license before they start operation, and (ii) because of the neglect of the closure cost may end up the project in negative balance at the end of the projects life (Gracia, 2008). A social license refers to the consent of the community to have the mine in the community. The social license is a non-permanent, intangible, informal approval by the community and therefore the company cannot take the acceptance of the community for granted and have to maintain the relation to allow the social license to survive.

14.2.2 Financial Assurance

The planning for closure as a whole life cycle strategy to manage environmental impacts of mining from its inception to end, i.e. from the development to closure stage depends on what is known as time factor. The higher the time gap between the occurrence of environmental damage and its remediation, generally the more resources will be required to resolve the problem, both human and financial (Warhurst & Noronha, 1999, p.7). To ensure that the reclamation and closure cost will be covered by the company many jurisdiction require guarantees for mine closure, i.e. financial assurance prior to mine opening.

14.2.3 Standards by International Financial Bodies

In the absence of regulatory agencies at the international level, financial institutions have emphasized the importance of closure for socially conscious and fiscally safe banking purposes. At present, 85 financial institutions have adopted the Equator Principles. The Equator Principles require that the financial institutions asses and manage social and environmental risk as part of the project financing and the borrower to conduct relevant social and environmental assessment to identify relevant social and environmental impacts and to ascertain risks of the proposed project. The International Finance Corporation (IFC) uses environmental and social screening criteria to recognize the magnitude of social and environmental impacts. The IFC's sustainability framework includes certain Performance Standards (PS), such as, (a) PS-1: Assessment and management of environment and social risks and impacts, (b) PS-6: Biodiversity conservation and sustainable management of living natural resources. Therefore integrated environmental management through planning for closure incorporates, (i) planning in anticipation of operations, and (ii) implementing this plan throughout the life of mine so that potential environmental effects associated with each production phase are avoided or minimized or managed (Warhurst & Noronha, 1999).

14.3 'RAT HOLE' COAL MINING IN MEGHALAYA

Meghalaya is the only state in India where coal mining is done privately given the Constitutional safeguards available to the community under the Sixth Schedule of exclusive right to private property with unlimited access to resources. The market driven artisanal ‗rat-hole' mining Meghalaya has shifted the livelihood from agro-based to mine based (Mukhopadhyay, n.d.). Tens of millions of people worldwide depend on artisanal mining for their livelihoods and incomes, far more than depend on large-scale mining. Artisanal mining tends to be most common in poor areas, magnifying its developmental implications and risks. Artisanal mining generates employment and income, but it is not always safe, well-monitored, legal or regulated. ASM activities can cause substantial negative environmental, health, and social impacts and its informal nature also can make artisanal mining an easy source of income for organized crime and armed conflicts.

Economic and Livelihood concerns rose due to the April 17, 2014 ban on ‒rat-hole' coal mining in Meghalaya by the National Green Tribunal (NGT) through an interim order following the petition filed by Assam based All Dimasa Student's Union and Dima Hasao District Committee for acidic discharge from coal mines of Meghalaya and pollution caused to Kopili River Downstream. The Unscientific mining of minerals poses a serious threat to the environment, resulting in reduction of forest cover and loss of biodiversity, soil erosion and pollution of air, water and land. The primitive and unscientific _rat-hole' method of mining adopted by private operators and related activities have caused large-scale environmental degradation and severe ecosystem destruction in the state. The ban has created a sense of economic insecurity and instability in the state which is likely to continue unless an environmentally sustainable scientific mining policy is adopted. The rat-hole coal mining in the state started as a small scale subsistence livelihood model for the families in the rural areas but over the time grew in volume and transformed into a full-fledged industry which at present has much potential to impact the local economy as well as satellite economy developed around the industry. A significant part of the coal extracted through rat-hole mining was exported to neighbouring Bangladesh through eleven land custom stations. The ban hit the thousands of brick kilns industries in the country (Bangladesh Brick Klins Hit by Ban on Meghalaya Coal Mining, 2015).

It is also noticed that crime rate is rising in the state which is otherwise regarded as a peaceful state. There are reports of rampant theft and incidents of loot in the mining area. Due to the unemployment and frustration many people are resorting to criminal activities and reports of kidnapping, rape and killings has increased several folds recently after the imposition of the NGT ban. There are vulnerability of creating sick and depressed human capital, broken families and bleak future. (Chauhan & Kharumnuid, 2016).There are immigrant workers engaged in the traditional mining, with the imposition of ban they are jobless, some are heading towards native place, some looking for alternative opportunities and some may end up involving in criminal activities.

The Meghalaya Mines & Mineral Policy 2012 is not yet been enforced as rules for concession Major Mineral are yet to be formulated. Despite the fact that the traditional unscientific method of rat-hole coal mining has adverse impact on the environment, the MMMP, 2012 seems in favour of continuation of the traditional rat-hole mining and exploitation of mineral resources in a scientific manner considering the interest of the state and its people.The estimated reserve of Coal in Meghalaya is 576.48 million tonnes. So, anguish in the affected stakeholder is to continue unless a proper scientific regulatory mechanism developed and the ban is withdrawn as there remains vast potential of exploiting these resources.

14.4 CONSTITUTIONAL SAFEGUARDS TO MEGHALAYA

There is another angle to the entire debate, as the existing central laws will come into conflict with the current customary practice in Meghalaya. As per the Coal Mines (Nationalization) Act 1973, all coal mines in the state belong to the Union Govt. and under the Mines and Minerals (Development and Regulation) Amendment Act, 2015 have provisions for allocation of Coal mines through auctions, which is contrary to the mining practice in the state. Under the clause (2) of Article 244- Administration of Scheduled and Tribal Areas' of the Constitution of India, provisions of Sixth Schedule shall apply to the tribal areas of Meghalaya. The Para One read with Para Twenty Part-II the Khasi Hills District, Jaintia Hills District and the Garo Hills District are autonomous districts.

The Para Three- Powers of the District Councils to make laws' the District Council for an autonomous district in respect of all areas within the district shall have the power to make laws with respect to:

a) The allotment, occupation or use, or setting apart, of land, other than any land which is a reserve forest for the purpose of agriculture or grazing or for residential or other non-agricultural purposes of for any other purpose likely to promote the interests of the inhabitants of any village or town. However the State Government may make compulsory acquisition of land, whether occupied or unoccupied for public purposes in accordance law in force for such acquisition.
b) The management of any forest not being a reserve forest
c) The use of any canal or water-course for the purpose for the purpose of agriculture
d) The regulation of the practice of jhum or other forms of shifting cultivation
e) The establishment of village or town committees or councils and their powers
f) Any other matter relating to village or town administration, including village or town police and public health and sanitation
g) The appointment or succession of Chiefs or Headmen
h) Marriage and divorce
i) Social customs

The Para Nine- Licenses or lease for the purpose of prospecting for, or extraction of, minerals the State Government shall grant license or lease for the purpose of prospecting for, or the extraction of, minerals. Royalties accruing each year from such license or lease shall be agreed upon by the Government and the District Court of such district. Any dispute arising out of such share of royalties shall be determined by the Governor of the State and shall be final.

As the primary objective of the Sixth Schedule is to protect, preserve and promote tribal culture and traditions, following the NGT imposing a blanket ban on mining and transportation of coal on April 17, 2014, the state Assembly had adopted a resolution urging the Centre to invoke Para 12A(b) of the Sixth Schedule through a presidential notification to ensure relevant provisions of two acts MMDR and Coal Mines Nationalisation Act, 1973, were exempted in Meghalaya (New Mining Policy Will Ensure Environment

Protection: Sangma, 2015). The clause (b) of Para 12A- 'Application of Acts of Parliament and of Legislature of the State of Meghalaya to autonomous districts and autonomous regions in the State of Meghalaya' of the Sixth Schedule empowers the President of India to direct through notification, that any Act of Parliament shall not apply to an autonomous district or regions in the State or shall apply to such district or region or any part thereof subject to such exceptions or modifications as he may specify in the notification and any such direction may be have retrospective effect.

14.5 MEGHALAYAN GOVERNMENT'S MEASURES

The ambitious Meghalaya Mines and Minerals Policy, 2012 (MMMP) (Meghalaya Mines and Mineral Policy, 2012) has been formulated with an aim to facilitate systematic, scientific and planned utilization of mineral resources and to streamline mineral based development of the State, keeping in view, protection of environment, land, health and safety of the people in and around the mining areas. The Policy will also dwell on ensuring optimal utilization of available mineral resources, realization of vast mineral potential, generate revenue for socio-economic development, uplift the economy of the State and enhance employment opportunities. The policy formulated with a focused mission of sustainable and eco-friendly growth of mineral deposits and mineral based industries with due regard to environment, conservation as well as upliftment of standards of living of the local people in and around the mineral bearing areas.

The Policy emphasises to secure proper linkage between exploitation of minerals and development of mineral industry, with preference to the local tribal people for development of deposits (MMMP, 2012, 1.3). The land tenure system in the State is largely governed by customary laws and practices. The Policy objectives will be in consonance with such laws and practices, as specifically stipulated in the Sixth Schedule to the Constitution of India and the Meghalaya Land Transfer (Regulation) Act, 1971and the relevant laws of the respective District Councils (MMMP, 2012, 1.5). The Policy also emphasises adoption of modern methods of mining would increase the safety of workers and reduce accidents as the antiquated and outdated method of manual extraction, which not only involves more time, labour and cost, but also constitute health hazard and risk to human life (MMMP, 2012, 1.6).

As the mechanized mining is expected to grow in the State gradually, which would obviously lead to increased demand for qualified technical personnel. Keeping this in mind the Policy realises the importance preparing the required Human Resource. Besides sponsoring students for degree courses in mining engineering and geology, the Policy advocates institutional arrangement in the State to prepare the local youth for acquiring statutory certificates from the Board of Mining Examinations/ Directorate General of Mines Safety so that the mine owners can be required to deploy the technically qualified personnel (MMMP, 2012,4.3). Accordingly initiatives will be taken to strengthen the ITI network in mining in order to equip employable local population with basic skills enabling absorption by the mining industry. Focus will be on excavation machinery management and repair, industry trades including welding, electrical repair, road and civil construction skills. Mining companies would be encouraged to adopt ITI in their area of operations for better absorption of skilled personnel (MMMP, 2012, 4.5). All Industries should ensure to provide maximum job opportunity to local people, particularly in skill labour and in clerical-staff jobs (MMMP, 2012, 20.e). In the mineral exploration small and traditional system of mining by local people in their own land shall not be unnecessarily disturbed (MMMP, 2012, 7.6). The state shall adopt scientific mining and which would be introduced in a phased manner for major minerals and for this, involvement of reputed institutions such as the Central Mining Research Institute, M/s Central Mine Planning & Design Institute (Coal India Ltd) would be sought (MMMP, 2012, 8.1.2). The Policy also recognises that the state has been endowed with the unique caves in the world and the Government shall make all endeavors to maintain and protect them (MMMP, 2012, 8.1.3). No mining activities will be allowed near rivers, river beds and streams to prevent pollution and ensure reclamation, rehabilitation and closure of mined out areas to000 facilitate environmental protection (MMMP, 2012, 8.1.4 & 9.1.v).

The Policy suggests adoption of appropriate measures to protect forests and maintain ecological balance in mining belts while pursuing mining activities. The Environment Management Plan shall adequately provide for controlling the environmental damage, restoration and reclamation of mined areas. Planting of trees and afforestation shall proceed concurrently with mineral exploitation, as far as practicable. Prevention and mitigation of adverse environmental effects due to mining operation, storage and processing of minerals, including disposal of mine-spoils shall form integral part of mining plan/ strategy in accordance with the norms and standards prescribed under the relevant Acts and Rules. Mitigation measures shall invariably include prevention and control of water pollution, gaseous pollutants, soil erosion and landslides, siltation, stabilization of waste dumping sites including repairing and re-vegetation of the affected forest area and land covered by trees (MMMP, 2012, 15).

Most importantly the Policy talks about Mine closure plan, which shall emphasize adequate post mining measures on restoration of mined landscapes, control of subsidence, control of pollution of surface and ground water especially from acid mine drainage and afforestation of the mined land and surrounding areas (MMMP, 2012, 15).Recognising the fact that mining activity needs to be done in a manner that does not permanently degrade the land, the State shall ensure that the mines in their mine closure plans make adequate provision for reclamation and/or restoration of the land to the best possible potential in collaboration with local communities, and for their use. Reclamation/restoration efforts shall specifically address issues of bringing land into productive use; reducing soil erosion through vegetative means; dealing with chemical pollutants of soil and water; improving the water regime and recharge potential; and mitigating the adverse visual impact.

Mine closure including progressive mine closure processes will be closely monitored and it will be ensured that stakeholders are taken into confidence at all stages through a transparent process facilitated by the State Government. Old and disused mines dating to prior to regulated Mine Closure shall be restored or rehabilitated using funds generated from royalties so as to enable local communities to regain the use of such lands. Land after closure shall be returned to the original landowners.

The Meghalaya Mineral Development Corporation Ltd. (MMDC) has also notified on June 2015 Appointment of Mining Consultant (individual or firms having personnel, meeting prescribed conditions) for carrying out mining operations in different parts of the state. Whereby, the consultant can provide services to MMDC and/or local miners/ mining operators to carry out scientific and technologically developed mining in the state. The qualification required for the position required a graduate in Mining Engineering with First Class Coal Mine Manager's certificate of competency from Directorate General of Mines Safety, Dhanbad (DGMS)[1] or Diploma in Mining Engineering with First Class Coal Mine Manager's certificate of competency from DGMS, Dhanbad along with 10 years and 15 years of experience respectively in senior level responsible position in underground/open cast coal mining.These experiences should be in mechanised mines using modern mining equipment (Appointment of Mining Consultant, 2015).

Initiatives and efforts by the state government give a clear indication that it is in a process of introducing state of the art technology and expertise in the mining sector generally and more specifically to the much talk about traditional coal mining. With the recent approval of the Meghalaya Minor Minerals Concession Rules in early July this year, to ensure protection of people engaged in mining of minor minerals and also considering the environmental aspect, it is also expected that the Concession Rules for Major mineral shall also be coming shortly.

14.6 REGULATION OF MINES IN INDIA

The history of mineral extraction in India can be traced back from the age of *Harappan Civilization*. Since India's independence there has been a significant growth in the mineral production, both in terms of quantity and value. The studies of various geologists suggest that India is a part of prehistoric land mass known as ‚Gondwanaland' to which Australia, South and Central Africa, and South America also a part. Therefore India should have similar types of minerals both in quantity and quality, unlike other countries belonging to ‗Gondwanaland' (National Mineral Policy, 2006, p. 10).

The Ministry of Mines, Government of India is responsible for survey and exploration of all minerals for mining and metallurgy except natural gases, petroleum and atomic minerals.It is the responsibility Indian Bureau of Mines to promoting conservation of mineral resources by way of inspection of mines, geological studies, scrutiny and approval of mining plans and mining schemes, conducting environmental studies and environment related activities, evolving technologies for upgradation of low grade ores and identifying avenues for their utilisation, preparation of feasibility reports for mining and beneficiation projects, preparation of minerals maps and National Mineral Inventory of minerals resources; providing technical consultancy services to mineral industry and functioning as a data bank for mines and minerals, and preparing of technical and statistical publications.

In furtherance to the economic reforms introduced by the Government of India during early 1990's, the National Mineral Policy of 1993 came into existence (Ministry of Mines, n.d.). Despite recognizing the need to encourage flow of private investment, including Foreign Direct Investment and introduction of state-of-art technology in exploration and mining (National Mineral Policy, 2006, p. i); the National Mineral Policy of 1993 also reflected a general sense of mineral scarcity in the country and therefore focused on mineral conservation and emphasised on the need to meet national and strategic requirements. However, the sense of mineral scarcity was replaced by a sense of mineral abundance by the National Mineral Policy of 2008.

In India mineral deposits are mostly located in remote and tribal areas of the country, where the primary source of livelihood depends on agricultural and forest produce (Trivedi, R., et. al., 2011, p. 31). Large extant of forest lands has been destroyed by mining and quarrying activities. The process of mineral exploration, development, production, and disposal of minerals affect the environment and ecology of the mined area and therefore mining has to be done in a way that causes least damage to natural resources such as air, water, soil, and biomass (National Mineral Policy, 2006, p. 66).

According to the National Environment Policy of 2006, it has to be ensured that provisions for environmental restoration, particularly mine closure in all approvals of mining plans and also institutionalise a system of post monitoring of such projects (Ministry of Environment and Forest and Climate Change, 2006). One of the major thrust areas in the 12th Five Year Plan was to develop an Action Plan by the State Directorates of Mining and Geology to create facilities for mine closure, sustainable mining practices and stakeholder protection (Ministry of Mines, 2011a).

14.7 LEGAL FRAMEWORK FOR MINING SECTOR

In India mines and mineral are managed by both Central Government and the State Governments under Entry 54 (Regulation of mines and mineral development to the extent to which such regulation and development under the control of the Union is declared by Parliament by law to be expedient in the public interest) of the Union List (List I) and Entry 23 (Regulation of mines and mineral development subject to the provisions of List I with respect to regulation and development under the control of the Union.) of the State List (List II) of the Constitution of India respectively. The combined reading of Entry 23 and 50 (Taxes on mineral rights subject to any limitations imposed by Parliament by law relating to mineral development) of List II makes it clear that, unless the Parliament (Central Government) makes any law in exercise of its powers in Entry 54, the powers of State Legislature will be exercisable under Entry 23 and 50. The Central Government enacted *The Mines Act, 1952* to regulate the working conditions in the mines and to provide certain amenities to the workers employed therein. The Central Government enacted *The Mines and Minerals (Development and Regulation) Act, 1957*, which incorporates the legal framework for regulation and development of mines and minerals in the country, except mineral oils (natural gas and petroleum). *The Mineral Concession Rules, 1960*, had been introduced to regulate grant of Reconnaissance Permits, Prospecting Licences and Mining Leases in respect of all minerals other than atomic minerals and minor minerals. Subsequently *The Mineral Conservation and Development Rules, 1988*was enacted for conservation and systematic development of minerals, except mineral oils (petroleum and natural-gas), coal, lignite, sand for stowing and minor minerals.

From the perspective of the present work it would be necessary to highlight specific legal framework relating to coal mining in India, in addition to general legal framework for mining. In this regard the most important is *The Coal Mine (Nationalisation) Act, 1973*, which deals with acquisition and transfer of the right, title and interest of the owners in respect of coal mines specified in the schedule to the Act, with a view to re-organising and re-constructing such coal mines so as to ensure rational co-ordinated and scientific development and utilization of Coal resources consistent with the growing requirement of the country, in order that the ownership and control of such resources are vested in the State and thereby so distributed as best to sub-serve the common good. In addition The Coal Bearing Areas (Acquisition and Development) Act, 1957; Coal Mines (Conservation and Development) Act, 1974; Coal Mines (Conservation and Development) Rules, 1975; The Colliery Control Rules, 2004; The Coal Mines (Special Provisions) Act, 2015; The Coal Mines (Special Provisions) Rules, 2014; The Auction by Competitive Bidding of Coal Mines Rules, 2012 also regulates the coal mining sector in the country.

The Ministry of Coal is responsible for development and exploitation of coal and lignite reserves in India. Coal Controller's Organization is a subordinate office of Ministry of Coal, Office of Coal Controller (earlier Coal Commissioner), established in 1916, is one of the oldest offices in Indian Coal sector. Main aim behind setting up this office was to have Government control to adequately meet the coal requirement during First World War. It works for effective control on production, distribution and pricing of coal. Mineral wealth is generated through combination of various proponents, primarily capital and labour, which is managed by the investor and the mining company in cooperation with the state, mineral rights owner and the affected communities. The depletable and non-renewable mineral resources can create substantial wealth, but for sustainable development these resources need to be managed, so that the wealth they generated can effectively compensate for depleting minerals (Rocha & Bristow, 1997).

However, in recent times the governments has realized the complex mixture of environmental, social, economic and developmental issues of mine closure and it is the responsibility of the government to ensure — (a) the mining industry has properly identified those issues and prepared to address any problems over the life of mining, and (b) the closure plan be implemented in consultation with all the major stakeholders like communities and government at all levels (Clark & Clark, 2005). The Government of India started to give a serious thought to the environmental degradation caused by mining sector in the early 2000. The government's effort to minimize the detrimental effects of mining can be inferred from the amendment made in 2003. In exercise of it powers conferred under Section 13 and Section 18 of *The Mines and Minerals (Development and Regulation) Act, 1957*; the Mineral Concession Rules 1960 and the Mineral Conservation and Development Rules 1988 were amended respectively time to time.

14.8 MINE & MINERALS CLOSURE PROVISIONS

According to the amendments every mine shall have two kinds of mine closure plan, such as (i) progressive mine closure plan, and (ii) final mine closure plan (The Mineral Conservation And Development Rules, 1988, Rule 23A). A progressive mine closure plan has to be submitted as a component of the mining plan, in case of fresh grant or renewal of mining lease and such plan shall be reviewed by the company in every five years' time. In case of an existing mine the company shall have to submit such progressive mine closure plan within 180 days of notification of such amendment (The Mineral Conservation And Development Rules, 1988, Rule 23B and The Mineral Concession Rules, 1960, Rule 22(5)(va)).A final mine closure plan has to be submitted for approval one year prior to the proposed closure of the mine (The Mineral Conservation And Development Rules, 1988, Rule 23C). The lessee shall not determine the lease in full or in part unless the final mine closure plan has not approved by the concerned authority and obtained a certificate for the same to the effect that proactive, reclamation and rehabilitations as per the approvals of the authority has been carried out by the lessee (The Mineral Concession Rules,

1960, Rule 29A). In case the lessee seeks any modification to the approved mine closure plan, have to submit the intended modification and explanation for such modification to the concerned authority (The Mineral Conservation And Development Rules, 1988, Rule 23D). The mining company will be responsible to ensure that proactive measures, reclamation and rehabilitation works have been carried out in accordance with the approved mine closure plan and shall submit a yearly report, stating the activity already done and reasons for any such failure thereof (The Mineral Conservation and Development Rules, 1988, Rule 23E).

The leaseholder has to furnish financial assurance before executing the mining lease deed to the concerned authority. The financial assurances could be in the following forms:

a) Letter of Credit from any Scheduled Bank,
b) Performance or surety bond,
c) Trust fund build up through annual contributions from the revenue generated by mine and based on expected amount sum required for abandonment of mines, or
d) Any other form of security or any other guarantees acceptable to the authority.

Any such financial assurance shall be liable to be released upon satisfactory completion of the mine closure plan and certification by the concerned authority (The Mineral Conservation And Development Rules, 1988, Rule 23F). As minerals like, mineral oils (petroleum and natural-gas), coal, lignite, sand for stowing and minor minerals is not covered by *The Mineral Conservation and Development Rules, 1988* the Government of India decided that all coal (including lignite) mining operation in India shall be governed by ‗Guidelines for Preparation of Mine Closure Plan' (MCP). Accordingly the Ministry of Coal, Government of India Notified the Guidelines on 27th August 2009. Since then the said Guideline for MCP has been amended and modified quite a few times until the recent notification dated 7th January 2013(Ministry of Coal, 2013).

All Coal (including lignite) mining operations shall prepare MCP, which shall be incorporated in the Project Report/Mining Plan for both new and existing mines. The MCP will have two components viz. i) Progressive or Concurrent MCP and ii) Final MCP. Therefore all coal (including lignite) mine owner shall adopt MCP duly approved by the competent authority. The competent authority for approval of MCP shall be the Standing Committee constituted by the Ministry of Coal for the purpose of approval of Mining Plans. However in the case of projects/mines of Government Companies, the competent authority to approve MCP will be the concerned Board of the Company wherever the power is so delegated for approval of Mining Plan.

Mining is to carried out in phased manner, thus a Progressive MCP would include various land use actives to be done continuously and sequentially during the entire period of the mining operations. Progressive mine closure shall be prepared for a period of every five years from the beginning of the mining operations. The MCP will be examined periodically in every five years period and to be subjected to third party monitoring for Assessment and Certification of works done in mine closure activities of Coal and lignite mines as per approved MCP, by agencies approved by Central Government like (Ministry of Coal, 2016):

a) Central Mine Planning and Design Institute Ltd. (CMPDIL).
b) National Environmental Engineering Research Institute (NEERI).
c) Indian School of Mines (ISM).
d) Indian Institute of Technology (IIT), Kharagpur.
e) Indian Institute of Engineering Science and Technology (IIEST), Shibpur.

Whereas the Final mine closure activities would start towards the end of mine life, and may continue even after the reserve are exhausted and /or mining is discontinued till the mining area is restored to an acceptable level by the Coal Controller as per the certification of agencies approved by Central Government. The Final MCP along with details of the updated cost estimates for various mine closure activities and Escrow Account set up at the beginning shall be submitted to the Ministry of Coal for approval at least five years before the intended final closure of the mine.

In the Closure Plan the mine owner has to provide mine description in respect of Geology describing the topography and general geology indicating rock types available, including toxic elements, if any, at the mine site; indicate the coal/lignite reserve available (proved, indicated and inferred) and the Mining Method to be followed to win the coal/lignite, mining machinery deployed, production level etc. The MCP shall describe the proposals/ measures to be implemented for reclamation (both physical and biological) and rehabilitation of mined out land including the manner in which the actual silt of the pit will be restored for post mining land use. All the proposals should be supported by relevant plans and sections depicting the method of land restoration reclamation/ rehabilitation and time frame for the same. In addition to this the MCP shall include plans for Water Quality management; Air Quality management; Waste management; Top soil Management; Coal beneficiation and Management of Coal Rejects; Infrastructure; Disposal of Mining Machinery; Safety and security; and Economic Repercussions of closure of mine.

Details of time schedule of all abandonment operations as proposed in the Closure plan should be described along with manpower and other resources required for the proposed plan. For Abandonment Cost, the cost is to be estimated (at the time of Mining Plan) based on activities such as barbed wire fencing all around the working area, dismantling of structures/ demolition and clearing of sites, rehabilitation of mining machinery, plantation, physical/biological reclamation, landscaping, biological reclamation of left out overburden dump, filling up of de-coaled void and most importantly post monitoring for 3 years, supervision charges for 3 years, power cost, protective and rehabilitation measures including their maintenances and monitoring. The estimated closure cost for an open cast mine will be around rupees six lakhs per hectare of the total project area and rupees one lakh per hectare for underground mining project (August 2009 price level) and these rates will be modified based on whole sale price index as notified by Government of India from time to time. The annual closure cost is to be computed considering the total project area as per the rate set by the Government and dividing the same by the entire life of the mine in years for new projects and balance life of mines in years for operating/existing mines and an amount equal to the annual cost is to be deposited each year throughout the mine life compounded at the rate of five percent annually. If the Mine owner fails to deposit the annual amount required to be deposited, the Government can withdraw the mining permission. For Financial Assurance the mining company shall open an Escrow Account with any Scheduled Bank, with the Coal Controller (on behalf of the Central Government) as exclusive beneficiary, prior to the permission is given for opening the mine by Coal Controller.

It is the responsibility of the Mine owner to ensure that the protective measures contained in the MCP including reclamation and rehabilitation works have been carried on in accordance with the approved MCP and the owner have to submit a yearly report before 1st July of every year to the Coal Controller any work carried out so far. Before surrendering the reclaimed land to the State Government concerned, the mine owner is required to obtain a Mine Closure Certificate from the Cole Controller to the effect that the protective, reclamation and rehabilitation measures has been carried out in accordance with the MCP.

The existing legal framework and the amendments in the recent past suggests that the government is giving a serious thought to the environmental, social and economic concerns of mine closure; keeping in mind the rapidly growing nature of the mining sector with the economic growth.

14.9 SUSTAINABLE DEVELOPMENT FRAMEWORK (SDF)

The National Mineral Policy of 2008 has recognized that mining is closely connected with forestry and environmental issues and a major part of the country's known mineral resources are in areas under forest cover. Although mining has the potential to disturb the ecological balance but considering the needs of the economic development, extraction of mining wealth becomes a priority. A comprehensive strategy for sustainable development has to be adopted, which can look after issues of bio-diversity and can ensure that mining activity does opt suitable measures to restore ecological balance. Therefore international best practices have to be adopted while dealing with the interested stakeholders and special attention will be given to protect the interest of host and indigenous communities. A framework for relief and rehabilitation will be adopted in line with National Rehabilitation and Resettlement Policy (Ministry of Mines, 2008, p. 2.3).

The Mining sector in the country has been facing severe criticism from different corners of the society relating to its performance and sustainable development concerns. The working definition for Sustainable Development Framework in the mining sector has been outlined as follows (Ministry of Mines, 2011b, p. 6):

> —Mining that is financially viable, socially responsible, environmentally, technically and scientifically sound; with a long-term view of development; uses mineral resources optimally; and, ensures sustainable post-closure land uses. Also one based on long-term, genuine, mutually beneficial partnerships between government, communities and miners, based on integrity, cooperation and transparency."

The Sustainable Development Framework will cover not only the important aspects that come under the preview of Ministry of Mines, but also other Ministries and departments as well. Therefore the Working Group set up by the Ministry of Mines shall consist of (Ministry of Mines, 2011b, p. 12):

a) Ministry of Environment and Forests
b) Ministry of Tribal Affairs
c) Federation of Indian Mining Industries
d) Indian Bureau of Mines
e) State Governments, represented by the Mines Secretary and Department of Mines and Geology.

The principles for incorporating sustainable development in mining sector are as such (Ministry of Mines, 2011b, p. 12):

a) Incorporating environmental and social sensitivities in decisions on leases.
b) Strategic assessment in key mining regions.
c) Managing impacts at the mine level through sound management systems.
d) Addressing land, resettlement and other social impacts.

e) Community engagement, benefit sharing and contribution to socio-economic development.
f) Mine closure and post-closure.
g) Assurance and reporting.

The Sustainable Development Framework is modeled on the lines of the International Council of Mining and Metals and International Union for the Conservation of Nature and Natural Resources, which will be used to regulate and assess the applications for mining leases, expansion/extension and green clearances (Suneja, 2012).

14.10 CONCLUSION

Despite the fact that India has a strong regulatory and institutional setup for environmental regulations, had to adopt integrated approach to regulate its mining sector. Though mining is century old practice in India, the volume of extraction was very negligible. But with the economic development and increase in domestic demand due to the development of the basic industries, i.e. power industry, cement industry, steel industry, etc., a multi-fold increase in the mining activity realised. According to the new National Mineral Policy of 2008, the mining industry is likely to grow further and more forest cover will be diverted for the same. With these diversions more forest dependent communities will be affected, both socially and economically. But for the national interest and to the greater interest of its citizens, mining has to go on. The most important part of the debate will be how India going to regulate the growing industry and its adverse impacts, especially on ecology and society. The concept of mine closure is just three decades old to the global mining sector and in an infant stage in India. Social licence issues, financial assurance and standards of international financial institutions, in a great extant contributed in developing the concept of mine closure through their inputs. However, the 2003 amendments to the Mineral Concession Rules 1960 and the Mineral Conservation and Development Rules 1988 and the subsequent adoption of Sustainable Development Framework are India's desperate attempt in this regard.

There are norms set by International best practice models and India as a responsible nation had already adopted various sustainable mining measures and is also in the process of developing more promising measures for Sustainable Development, which are continuously evolving. But in the context of mining in Northeast India and more specifically Meghalaya the recent NGT ban on rat hole mining is a great cause economic instability in the state as well as neighboring states. The unemployment of mine labourers had direct impact on rise in crime and the mine owners are also experiencing economic losses. Due to policy lack to tackle unscientific and illegal mining in Meghalaya may further give rise to Resource Curse in the region. Therefore it is very much certain that the Sustainable Mining norms set by regulatory agencies are going to clash with the existing mining practices in the state. It is always difficult for a community, which is traditionally engaged in an unscientific practice to adapt to a scientific and sustainable practice. However believing on the concept of Sustainable Development we all together have to face these transitory economic and livelihood concerns. It is also true that Multidimensional integrated approaches by Union and State Governments, NGOs can provide livelihood support through training and capacity building etc., by engaging the affected communities and various other stakeholders.

At this stage no conclusion can be drawn regarding the effectiveness of policies. Mining processes are time taking processes, mostly continues for many decades and only at the end of the mine life cycle an evaluation could be possible. The mining sector is in the rise in India and the policies of prospecting mine closure plan and post closure along with other environmental jurisprudence will help the mining sector to be sustainable sector. Keeping in view the approach for sustainability in mining sector there is more stringent provisions likely to come under the preview of legal framework in the years to come.

REFERENCE

1. Appointment of Mining Consultant. (2015, June 17). Meghalaya Mineral Development Corporation Ltd. Ref No. MMDC/MMMP/2014-5578. Retrieved from http://megdmg.gov.in/pdf/MMDC%20appointment%20of%20mining%20consultant.pdf
2. Bangladesh Brick Klins Hit by Ban on Meghalaya Coal Mining. (2015, January 12). *Zee News,* Retrieved from

http://zeenews.india.com/news/south-asia/bangladesh-brick-kilns-hit-by-ban-on-meghalaya-coal-mining_1529011.html
3. Chauhan, K & Kharumnuid, I. (2016, August 5). Institutional Policy and Its Role in Sustainable Resource Management and Development: A Critical Analysis of the _NGT' Ban on Rat-Hole Mining in Meghalaya, India. *SSRN*. Retrieved from https://papers.ssrn.com/sol3/papers.cfm?abstract_id=2819025
4. Clark, A. L. & Clark, J. C. (2005, June). An International Overview of Legal Framework For Mine Closure. *International Development Research Center (IDRC)*. Retrieved from http://www.elaw.org/node/3715
5. Hodge, R. A. & Killam, R. (2003, December 7). Post Mining Regeneration Best Practice Review: North American Perspective. In Report *Prepared for ECUS Environmental Consultancy; University of Sheffield, England and The Eden Project*. Retrieved from http://www.anthonyhodge.ca/publications/Post_Mining_Regeneration.pdf
6. Hoskin, W. M. A. (2005). Mine Closure: the 21st Century Approach – Avoiding Future Abandoned Mines. In *International and Comparative Mineral Law & Policy: Trends and Prospects*. Bastida, et al., London: Kluwer Law International
7. Garcia, D. H. (2008). *Overview of International Mine Closure Guidelines*, Paper presented at the 2008 Meeting of the American Institute of Professional Geologists, Arizona Hydrological Society, and 3rd International Professional Geology Conference, Flagstaff, Arizona, USA, September 20-24, 2008. Published by American Institute of Professional Geologists. Retrieved from http://www.srk.co.uk/files/File/papers/Mine-Closure-Guidelines.pdf
8. The Meghalaya Mines and Mineral Policy. (2012). The Gazette of Meghalaya, Mining and Geology Department, Govt. of Meghalaya. No.MG.40/2010/200. November 5th 2012. Shillong. Retrieved from http://megdmg.gov.in/pdf/Extra%20Ordinary%20Gazette%20Mines%20and%20Menirals%20Policy%202012.pdf
9. Mining for the Future. (2002, April). *Appendix B: Mine Closure- Working Paper*. In *Report Prepared by MMSD. Project of IIED*. No.34. Retrieved from http://pubs.iied.org/pdfs/G00884.pdf
10. Ministry of Coal. (2013, January). *Guidelines for Preparation on Mine Closure Plan-Reg*. Government of India. No.55011-01-2009-CPAM. Retrieved from http://www.cmpdi.co.in/env/MCP%207_1_2013.pdf
11. Ministry of Coal. (2016, June). Addition of IIT, Khargpur and IIEST, Shibpur as agencies for Assessment and Certification of works done of Mine Closure Activities of coal and lignite mines as per approved Mine Closure Plan (Progressive & Final) –Reg. Government of India. No.55011-01-2009-CPAM. Retrieved from http://coal.nic.in/sites/upload_files/coal/files/curentnotices/29-06-2016.pdf

12. Ministry of Environment and Forest and Climate Change. (2006). *National Environment Policy, 2006.* Government of India. Retrieved from http://envfor.nic.in/sites/default/files/introduction-nep2006e.pdf
13. Ministry of Mines. (n.d.). *Mining Policy Legislation (Overview).* Government of India. Retrieved from http://mines.nic.in/UserView?mid=1319
14. Ministry of Mines. (2008). *National Mineral Policy, 2008 (For Non-Fuel and Non-Coal Minerals).* Government of India. Retrieved from http://mines.nic.in/writereaddata/Content/88753b05_NMP2008[1].pdf
15. Ministry of Mines. (2011a, September). *Report of The Working Group on Mineral Exploration and Development (Other than Coal and Lignite) for The 12th Five Year Plan, Sub Group-I.* Government of India. Retrieved from http://mines.nic.in/writereaddata/Contentlinks/6b0b0e9c0fa540c9a51801df96697531.pdf
16. Ministry of Mines. (2011b). *Sustainable Development Framework (SDF) for Indian Mining.* 30th November. Retrieved from http://mines.nic.in/writereaddata/filelinks/2155afeb_FINAL%20REPORT%20SDF%2029Nov11.pdf
17. Morrey, D. R. (1999). Integrated Planning for Environmental Management during Mining Operations and Mine Closure. *Minerals & Energy.* Vol.14 No.4
18. Mukhopadyay, L. (n.d.). *Sustainable Development – A Path Dependent Analysis to the Rat hole Coal Mining in Jaintia Hills District, India.* Retrieved from http://economics.ucr.edu/repec/ucr/wpaper/13-06.pdf
19. National Mineral Policy. (2006). *Report of the High Level Committee.* Planning Commission, Government of India. December 2006. New Delhi. Retrieved from http://mines.nic.in/writereaddata/Filelinks/46ff58f0_rep_nmp.pdf
20. New Mining Policy Will Ensure Environment Protection: Sangma. (2015, March 15). *The Times of India, Shillong.* Retrived from http://timesofindia.indiatimes.com/city/shillong/New-mining-policy-will-ensure-environment-protection-Sangma/articleshow/51412035.cms
21. Otto, J. M. (2009). Global Trends in Mine Reclamation and Closure. In*Mining, Society and A Sustainable World.* Richards, J. P., ed., Germany: Springer
22. Rocha, J. & Bristow, J. (1997). Mine Downscaling and Closure: An Integral Part of Sustainable Development. *Journal of Mineral Policy, Business and Environment Raw Materials Report.* Vol.12 No.4
23. Sessoon, M. (2000). Environmental Aspects of Mine Closure. In *Mine Closure and Sustainable Development.* 116, Khanna, T., ed., London: Mining Journal Books Ltd.
24. Suneja, K. (2012, August 21). Sustainable Development Framework to Complement Green Nod for Mining. *The Financial Express.* New Delhi.

Retrieved from http://www.financialexpress.com/news/sustainable-development-framework-to-complement-green-nod-for-mining/990803
25. Trivedi, R., Sangode, A. G., Chakraborty, M. K. & Soliria, M. R. (2011).A Progressive Mine Closure Plan for Semi-mechanized Quarry of Dolomite. *The Indian Mining and Engineering Journal*. February. Retrieved from http://cimfr.csircentral.net/279/1/scan1.pdf
26. Warhurst, A. & Noronha, L. (1999). Integrated Environmental Management and Planning for Closure: The Challenges. *Minerals & Energy*. Vol.14 No.4
27. World Bank & International Finance Corporation. (2002). It's Not Over When It's Over: Mine Closure Around the World. In *Mining And Development-Global Mining*.Retrieved from http://siteresources.worldbank.org/INTOGMC/Resources/notoverwhenover.pdf

---

[1]DGMS, Govt. of India is the Regulatory Agency under the Ministry of labour and employment, Government of India in matters pertaining to occupational safety, health and welfare of persons employed in mines (Coal, Metalliferous and oil-mines).

Chapter 15

Environmental Impact Assessment of Mining in India: A Review of Legal and Institutional Mechanism

Aijaj Ahmed Raj & Zubair Ahmed

Abstract: *The particular geological condition determines the method of mining in certain region. Indiscriminate mining has always been a great contributor in the degradation of natural resources and the destruction of habitat thus poses a great threat to the biodiversity. Untreated waste material of huge quantities is being released by several mining agencies in the mining regions. If proper care is not taken for waste disposal, it will degrade the surrounding environment irreparably. Mining industries in India has always a tendency to undermine the role of EIA in its pre-operational, during mining and post operational plans.*

The basic law governing the mines is The Mines Act, 1952. However, most of the EIA related provisions of the Act do not provide for precautionary measures, but post-EIA monitoring i.e., remedial measures. Though a comprehensive EIA is not mandatory in the Act, it makes elaborate provisions for disaster mitigation. Section 57 of The Mines Act, 1952, provides that the EIA principles can be institutionalised under the regulatory and rule making power of the Central Government. The 1987 Amendment to the Act now contains a specific chapter on mines and environment. This paper throws light on the burning issues of mining, their impact on the environment and the laws governing the mining activities in India. The importance of conducting suitable assessment studies beforehand, to learn the potential adverse impact of mining on the ecology and the environment has also been emphasised.

Keywords: *Environmental Impact Assessment, Environmental Impact Assessment of Mining, Environment Management Plan, EIA Process, EIA Notification and EIA Legislation*

15.1 INTRODUCTION

Mining has been one of the important economic activities throughout the world. The access and use of the most of our energy resources have been possible through mining. Through mining we have the access to the rough materials for the society. Mining activities produce large quantities of waste and often they are not taken care of which can have deleterious impacts on the surface of the earth and can pose a great threat to the environment. The exploitation and valorification of the mineral resources results in the environmental perturbation with large scale ramifications. Even though mining is done in a particular area of a greater landmass, its impacts on environment extend to much greater area on the environment, not only on public health, but also to all the flora and fauna of that greater landmass. Mining is a long process starting from exploration of minerals, exploitation of the same and processing and at last finally the finished product comes to the market available for the consumer. In this whole long process of mining, great threat involves to the environment at every level. Lack of proper planning and negligence of regulations, mining activities cause irreparable damage, degradation and great harm to the environment and ecology.[1]

Mining activity puts tremendous pressure on local flora and fauna; particularly where division of forest land for mining takes place. Mining adversely affects the environment as it destroys vegetation, causes extensive soil erosion and alters microbial communities. However, the contributions of mining towards the economic development of the nation cannot be undermined although it also has a great impact upon human health. The mining activities also leave negative impacts on the natural environment particularly in the

---

[1] Ahmad, A. F. *et. al.*(2014). Impact of Mining Activities on Various Environmental Attributes with Specific Reference to Health Impacts in Shatabdipuram, Gwalior, India, *IRJES*, 3(6), 81.

cultivatable land and natural forest area.[2] Thus, a holistic approach to mining activities, keeping in mind the concerns regarding the local habitats and ecosystem, is necessary.[3]

The paper discusses environmental impact management of mining, the legal and institutional mechanism, the issues and challenges in the proper implementations of the environmental impact assessment in India. The paper also examines regional policies and the local initiatives in the recent years in the mining sector and their effectiveness in protecting environment at local level. It has also brought forth the issues that are yet to be resolved for more effective environmental impact management of in India.[4]

15.2 ENVIRONMENTAL IMPACT ASSESSMENT

The environmental impact assessment (EIA) process is an interdisciplinary and multistep procedure to ensure that environmental considerations are included in decisions regarding projects that may impact the environment. EIA can be defined as the study to predict the effect of a proposed activity or project on the environment. A decision making tool, EIA compares various alternatives for a project and seeks to identify the one which represents the best combination of economic and environmental costs and benefits. The purpose of the EIA process is to inform decision-makers and the public of the environmental consequences of implementing a proposed project. The EIA document itself is a technical tool that identifies, predicts, and analyzes impacts on the physical environment, as well as social, cultural, and health impacts. If the EIA process is successful, it identifies alternatives and mitigation measures to reduce the environmental impact of a proposed project. The EIA process also serves an important

---

[2] Guha, D. (2014). A Case Study on the effects of Coal Mining in the Environment Particularly in Relation to Soil, Water and Air causing a socio economic Hazard in Asansol-Raniganj Area, India *IRJSS,* 3(8), 39.
[3] Goswami, S. Impact of Coal Mining on Environment: A Study of Raniganj and Jharia Coal Field in India.
[4] Saviour, M.N. (2012). Environmental Impact of Soil and Sand Mining: A Review, *IJSET,* 1(3), 126.

procedural role in the overall decision-making process by promoting transparency and public involvement.

EIA categorically examines both beneficial and adverse consequences of a project and ensures that these effects are taken into account in the planning of the project. It helps in identifying possible environmental effects of the proposed project, proposes measures to mitigate adverse effects and predicts whether there will be significant adverse environmental effects, even after the mitigation is implemented.

Many countries now have laws stipulating that unless an EIA study is carried out (particularly for large infrastructure projects), permission for construction will not be granted by the local authority. The educational one is equally important and probably a forerunner to the legal role to educate everyone, one involved professionals and users included, of the potential environmental impacts of anything we do. EIA is a tool used to identify the environmental, social and economic impacts of a project prior to decision - making. It aims to predict environmental impacts at an early stage in project planning and design, find ways and means to reduce adverse impacts, shape projects to suit the local environment and present the predictions and options to decision makers. By using EIA both environmental and economic benefits can be achieved, such as reduced cost and time of project implementation and design, avoided treatment/cleanup costs and impacts of laws and regulations.

15.3 EIA IN INDIA

Indian experienced EIA around 20 years back. It was for the first time in 1976-77 when the Planning Commission asked the Department of Science and Technology to examine the river valley projects from an environmental angle. This was subsequently extended to cover those projects, which required the approval of the Public Investment Board. Till 1994, environmental clearance from the Central Government was merely an administrative decision and there was no legislation for the same. In 1994, the Union Ministry of Environment and Forests (MoEF), Government of India, under the *Environmental (Protection) Act* 1986, promulgated an EIA notification and makes Environmental Clearance (EC) mandatory for expansion or modernisation of any activity or for setting up new projects listed in Schedule 1 of the notification. Since then there have been 12 amendments made in the *EIA notification* of 1994. The MoEF recently notified new EIA legislation in September 2006. The notification makes it mandatory for various projects such as mining, thermal power plants, river valley, infrastructure (road, highway, ports, harbours and airports) and industries including very small electroplating or foundry units to get environment clearance.

EC respectively from MoEF or the concerned State Environmental Impact Assessment Authorities (SEIAAs). Where state level authorities have not been constituted, the clearance would be provided by the MoEF. Further, the notification provides for screening (determining whether or not the project or activity requires further environmental studies for preparation of EIA), scoping (determining the detailed and comprehensive Terms of Reference (TOR), addressing all relevant environmental concerns /questions for the preparation of an EIA Report), public consultation (ascertaining concerns of affected persons) and appraisal of project proposals (based on the public consultations and final EIA report). EC is required in respect of all new projects or activities listed in the Schedule to the 2006 notification and their expansion and modernization, including any change in product - mix.

The amendments to *EIA Notification* of 1st December 2009 exempts environmental clearance process the biomass based power plants up to 15 MW, power plants based on non-hazardous municipal solid waste and power plants based on waste heat recovery boilers without using auxiliary fuel.[5]

15.4 EIA & MINING LAWS

The Mines Act is laced with the singular object of extraction of mineral resources and current safeguard in relation to labour welfare. The EIA angle has been grossly overlooked in the basic law governing the mines. In general, resources in India are jointly managed by central and state government. The proprietary title vests

---

[5] Surjith, K. (2010). Environment Impact Assessment in India, Retrieved from: http://www.ies.gov.in/myaccountprofileview. php?memid=399

in the federating states while the centre has jurisdiction over mines and minerals development. *Mines and Minerals (Regulation and Development) Act* (MMRDA) was enacted in 1957. With regard to mining, *MMRDA* 1957 and *Mines Act* 1952 are the main piece of legislations on mining in India. *Forest Conservation Act* 1980 and *Environment Protection Act* 1986 enacted for the protection of forest and environment go together with those legislations and are also applicable to mining.

It was in later phase that in view of the dubious record of the mining industries in undermining the social ecological imperatives, the Mines and Mineral (Development and Regulation) Rule, 1987 has specifically engrafted provisions of pre and post EIA. The *MMRDA* and other legislations are guided by the overall *National Mineral Policy* (NMP) of the Government of India, which was first outlined in 1993 and then revised in 2002 and again in 2008 based on the recommendations of *Huda Committee*. Thus, to give effect to the new *National Mineral Policy*, 2008, the *MMDRA* has been amended to ensure that the development of mineral resources is in accordance with the national policy goals.[6] Mining industries in India has a dubious record in undermining the social and ecological imperatives. The basic law governing the mines makes a peripheral reference to EIA under the safety provisions.[7]

The EIA principles can be institutionalized under the regulatory and rule making power of central Government. The Central Government may, make regulations for providing for the safety of the persons employed in a mine their means of entrance there into and exit there from, the number of shifts or outlets to be furnished, and the fencing of shafts, pits outlets pathways and subsidence. The safety of passage assured by providing for the safety of the roads and working places in mines, including the sitting, maintenance and extraction or reduction of pillars or blocks of minerals and the maintenance of sufficient baniers between mine and mine. The Central Government can also provide safeguards against explosion or ignitions of

---

[6] Khanna, A. A. (2013). Governance in Coal Mining: Issues and Challenges, *TERI-NFA*, 8-9.

[7] Nomani, Z. M. (2010). *Environment Impact Assessment Laws.* New Delhi, Satyam Law International.

inflammable gas or dust or accumulations of water in mine and against danger arising there from and for prohibiting, restricting or regulating the extraction of minerals in circumstances likely to result in the premature collapse or working or to result in or to aggravate the collapse of working or irruptions of water or ignitions in mines. Under the rule making power the central government can prescribe the standard of sanitation to be maintained. The 1987 Amendment to the Act now contains a specific chapter on mines and environment. These provisions can well take care of the EIA principles and policies.[8]

15.4.1 EIA of Mining under Central Laws
The mining sector is guided is principally managed by the Ministry of Mines, Ministry of Environment and Forest, Ministry of Coal, and Ministry of Labour. Ministry of Mines (MoM) looks after the developing policies and strategies of mining. The bodies responsible for formulating legislations to mitigate and control environmental pollution and planning under the administrative control of MoC are Coal Controller, various committees, the Coal Mines Provident Fund Organization, and the Commissioner of Payments Office. They also work for promotion and coordination of environmental programs. Regional offices of MoEF and Central Pollution Control Board (CPCB) assist the MoEF in executing such responsibilities. Likewise, in Ministry of Labor (MoL), Directorate General of Mines Safety (DGMS) is the authority to look after the enforcement of the statutory provisions on safety, health, and welfare in the workplace.[9] Some other ministries though they do not have any direct regulatory responsibility towards the mining sector but may take decisions that have the potential to impact the competitiveness of coal industry. These include Ministry of Railways (MoR), Ministry of Surface Transportation (MoST), Ministry of Finance (MoF), Ministry of Power (MoP), Ministry of Industry (MoI) and Ministry of Steel (MoS) etc.

---

[8] Nomani, Z. M. (2004). *Natural Resource Laws & Policy*. New Delhi, Uppal Book Publishing.
[9] Khanna, A. A. (2013). Governance in Coal Mining: Issues and Challenges, *TERI-NFA*, 10.

15.4.2 EIA of Mining under State

In the states, the departments like the Department for mining, Department for forest, Department for environment, and State Pollution Control Board (SPCB)s look after the mining activities within their jurisdiction. The Department for Mining has the responsibility of reviewing of applications for mineral titles, supervising their compliance with the standard guidelines and collecting the data. Another important role is played by the Department of Forests in granting forest clearances and prescribing compensatory a forestation. Department of Environment mostly functions at the lower level, because of lack of skills which are indispensable for policy planning and implementation of the same and thus they become bound to restrict themselves with the routine budgetary functions for State Pollution Control Boards (SPCBs). By virtue of their traditional role, SPCBs are larger institutions which were constituted to implement the Water Act in the states. Their main functions are ensuring implementation of the provision of relevant Acts; laying down, modifying or annulling effluent and emission standards; planning and execution of programs for prevention, control, or abatement of pollution and advising state governments on the same.

15.4.3 EIA of Mining at Local Level

Municipalities and Panchayats are the bodies at local level, which to play a very significant role in environmental protection and management at the district level. Their roles and responsibilities include, among others, soil conservation, land improvement, and management of conjunctive use of resources such as water. In addition to them, some other institutes like district collector or magistrates, department responsible for collecting taxes and royalties, and department responsible for issuing licenses for mining operations perform indirect regulatory functions.[10]

---

[10] Khanna, A. A. (2013). Governance in Coal Mining: Issues and Challenges, *TERI-NFA*, 11.

15.5 EIA PROCEDURE IN INDIA

The whole EIA process is the final outcome of the following phases:

1. *Proposal for the Project:* Any proponent embarking on any major development project shall notify Impact Assessment Agency (IAA) in writing by the submission of a project proposal. The project proposal shall have to provide with all relevant information which includes a land use map in order for it to move to the next stage which is screening. The submission of a project proposal is the first stage which signifies the commencement of the EIA process.

2. *Screening:* The second phase is screening. In screening the necessity a project is examined to see whether environmental clearance is required as per the statutory norms. At this stage, the type of project is decided and the need for Environmental Clearance found out. If Environmental Clearance is prescribed, the proponent may consult IAA.

3. *Scoping and Consideration of Alternatives:* Scoping is a process of detailing the terms of reference of EIA. It is done by the consultant in consultation with the project proponent and guidance, if need be, from IAA. The Ministry of Environment and Forests provides clear guidelines for different sectors, which outlines the significant issues to be addressed in the EIA studies. Quantifiable impacts are to be assessed on the basis of magnitude, prevalence, frequency and duration and non-quantifiable impacts, significance are commonly determined through the socioeconomic criteria.

4. *Base Line Data Collection:* Base line data is the existing environmental status of the particular area. The site specific primary data is to be monitored for the identified parameters and supplemented by secondary data if available.

5. *Impact Prediction and Assessment Of Alternatives:* Impact prediction is a process in which the environmental consequences of a particular project is examined and try to find out its alternatives. For every project, possible alternatives are tried to be identified and their environmental impacts are compared. Then a comparative analysis of alternatives is made and the best alternative is selected. A mitigation plan is drawn up for the selected option and is supplemented with an Environmental Management Plan (EMP) to guide the proponent towards environmental improvements.

6. *EIA Report:* An EIA report should provide clear information to the decision maker on the different environmental scenarios without the project, with the project and with project alternatives. The proponent prepares detailed Project report and provides information in logical and transparent manner. The IAA then examines whether the procedures have been followed as per MoEF notifications.

7. *Public Hearing:* The law mandates that the public must be informed and consulted on a proposed development project. Before the proposals are sent to MoEF for obtaining environmental clearance, the State Pollution Control Boards will conduct the public hearing. Any one likely to be affected by the proposed project has access to the Executive Summary of the EIA. The affected persons may be: a) Bonafide local residents; b) Local associations; c) Environmental groups: active in the area; d) Any other person located at the project sites of displacement. They are to be given an opportunity to make oral/written suggestions to the State Pollution Control Board as per Schedule IV.

8. *Decision Making*: Decision making process includes consultation between the project proponent (assisted by a consultant) and the impact assessment authority (assisted by an expert group if necessary). The decision on environmental clearance is arrived at through a number of steps including evaluation of EIA and EMP.

9. *Monitoring The Clearance Conditions*: Monitoring has to be done during both construction and operation phases of a project. It is done not just to ensure that the commitments made are complied with but also to observe whether the predictions made in the EIA reports are correct or not. Where the impacts exceed the predicted levels, corrective action should be taken. Monitoring also enables the regulatory agency to review the validity of predictions and the conditions of implementation of the Environmental Management Plan (EMP). The Project Proponent, IAA and Pollution Control Boards should monitor the implementation of conditions. The proponent is required to file once in six months a report demonstrating the compliance to IAA.[11]

15.6 ISSUES AND CHALLENGES IN EIA OF MINING

Even after having policy and legislative framework, mining in India has been continuously affecting environmental condition in and around the mining areas to deteriorate over the years. The behaviour of the mining companies with regard to mine closure and restoration shows their clear violations to the laws and policies as mandated by the same, and the sad part is that hardly there have been any stringent actions taken against such violations.

Points of major challenges in the EIA of mining are:

a) Inaccurate and incomplete data & non-adherence of scientific methodology
b) There is a conflict of interest as the project proponents gets the EIA conducted
c) Fabrication, repetition of old information, or avoiding crucial rather more important facts
d) Public hearings are often preposterous
e) Most often smaller areas (less than 5 hectares) are excluded from the requirement of an EIA, leading to unchecked mining and exploitation
f) Assessments are conducted tactfully during summer season when the land is drier and meaner. As a result, water courses are generally neglected in EIAs.
g) Poor implementation and lack of inappropriate monitoring have left the EIA process as a mere administrative formality

The lack of a proper mechanism for co-ordination across various government departments and institutions makes it problematic in the proper implementations of environmental impact assessment norms. Apart from co-ordination issues, overlapping of jurisdictions have been observed, which again creates problems in enforcement and

---

[11] (2016). Retrieved from: http://coe.mse.ac.in/EIAprocedure.asp

implementation of regulations. Regional office of MoEF and SPCBs, for instance, have similar roles and responsibilities of monitoring and enforcing various laws applicable to air, water, and land. The regulatory bodies are clearly ineffective in regulating and monitoring the different aspects of coal development. Shortages of skilled manpower and inadequate availability of equipment are two major factors for their ineffectiveness. These factors were highlighted in discussions with various regulatory bodies that include SPCBs, DGMS, State transport department, State forest department etc. Along with these, political influence has always been found as one of the major factors behind no or inadequate responsiveness of the regulatory bodies to the observed fallacies. In many cases, loopholes in implementations are deliberately ignored given the importance of coal for electricity generation and the grave impact on the economy as a result of any disruptions in the coal supply.[12]

15.7 CONCLUDING OBSERVATIONS

Mining shares a very important contribution on the economic, social and environmental fabric of a much greater area adjoining to the mining. Although the economic contributions of mining activities cannot be undermined, the land degradation due to mining and the ecological and socio-economic problems resulting out of such operations are really very dangerous. Apart from these, mining has always been responsible for huge social costs in the form of displacement, loss of livelihood and social exclusion. And it is compounded due to the dominance of open cast mining which needs larger tracts of land which ultimately results in larger loss of habitats and livelihoods. The affected are not only the people who have lost their land and houses, but also the people living in the much larger area surrounding the mining projects. Villages surrounding the projects, even though not displaced in many instances, have to suffer from degraded environmental hazards like water scarcity, air, noise, and water pollution, health impacts etc. So, it is very important to conduct suitable assessments to examine the potential adverse impact of mining on flora and fauna of a larger area before allowing

---

[12] Guha, D. (2014). A Case Study on the effects of Coal Mining in the Environment Particularly in Relation to Soil, Water and Air causing a socio economic Hazard in Asansol-Raniganj Area, India, *IRJSS*, 3(8), 28.

the activities of mining. The adverse impacts should be identified at the planning stage itself so that preventive measures may be taken to evade them. For this, one should be aware of the various activities affecting the environment.

REFERENCES

1. Ahmad, A. F. et. al.(2014). Impact of Mining Activities on Various Environmental Attributes with Specific Reference to Health Impacts in Shatabdipuram, Gwalior, India, *IRJES*, 3(6), 81.
2. Guha, D. (2014). A Case Study on the effects of Coal Mining in the Environment Particularly in Relation to Soil, Water and Air causing a socio economic Hazard in Asansol-Raniganj Area, India *IRJSS*, 3(8), 39.
3. Goswami, S. Impact of Coal Mining on Environment: A Study of Raniganj and Jharia Coal Field in India.
4. Saviour, M.N. (2012). Environmental Impact of Soil and Sand Mining: A Review, *IJSET*, 1(3), 126.
5. Surjith, K. (2010). Environment Impact Assessment in India, Retrieved from: http://www.ies.gov.in/myaccountprofileview. php?memid=399
6. Khanna, A. A. (2013). Governance in Coal Mining: Issues and Challenges, *TERI-NFA*, 8-9.
7. Nomani, Z. M. (2010). *Environment Impact Assessment Laws.* New Delhi, Satyam Law International.
8. Nomani, Z. M. (2004). *Natural Resource Laws & Policy.* New Delhi, Uppal Book Publishing.
9. Khanna, A. A. (2013). Governance in Coal Mining: Issues and Challenges, *TERI-NFA*, 10.
10. Khanna, A. A. (2013). Governance in Coal Mining: Issues and Challenges, *TERI-NFA*, 11.
11. (2016). Retrieved from: http://coe.mse.ac.in/EIAprocedure.asp
12. Guha, D. (2014). A Case Study on the effects of Coal Mining in the Environment Particularly in Relation to Soil, Water and Air causing a socio economic Hazard in Asansol-Raniganj Area, India, *IRJSS*, 3(8), 28.